T0339699

Bostrichidae (Coleoptera)

World Catalogue of Insects

VOLUME 17

The titles published in this series are listed at *brill.com/wci*

Bostrichidae (Coleoptera)

By

Petr Zahradník
Jiří Háva

BRILL

LEIDEN | BOSTON

Cover illustration: *Bostrichus capucinus* (Linnaeus, 1758). Photo by Pavel Krásenský.

The Library of Congress Cataloging-in-Publication Data is available online at https://catalog.loc.gov
LC record available at https://lccn.loc.gov/2024946851

Typeface for the Latin, Greek, and Cyrillic scripts: "Brill". See and download: brill.com/brill-typeface.

ISSN 1398-8700
ISBN 978-90-04-70790-0 (hardback)
ISBN 978-90-04-70791-7 (e-book)
DOI 10.1163/9789004707917

This book is printed on acid-free paper and produced in a sustainable manner.

This catalogue is dedicated to the *Pierre Lesne* (**9. 4. 1871–†10. 11. 1949*),
the famous expert of the family Bostrichidae

∵

Pierre Lesne (1871–1949)

Contents

Acknowledgements

For critically reading the manuscript and providing valuable comments and additions, and technical help with literature, we are indebted to the following colleagues: A. Bezděk (Czech Republic), L. G. Bezark (U.S.A.), R. A. Beaver (Thailand), J. Borowski (Poland), M. Knížek (Czech Republic), M. Kučerová (Czech Republic), A. Legalov (Russia), L. Y. Liu (Taiwan), P. Mikulčíková (Czech Republic), G. Nardi (Italy), J. Pouserreau (France), K. Šimerová (Czech Republic), Y-F. Zhang (China).

The book was supported by the Ministry of Agriculture of the Czech Republic, institutional support MZE-RO0118.

Introduction

The Bostrichidae family is one of three families of the Bostrichoidea superfamily (together with the Dermestidae and Ptinidae). Its distribution is global; the centre of its presence can be found in tropical zones with gradually decreasing distribution of species southward and northward.

This publication is based on the world catalogue (Borowski & Węgrzynowicz 2007a) modified by Borowski (2020) and T. Borowski (2020, 2021) which summarizes findings by numerous authors since the establishment of the binominal system (Linnaeus, 1758). Leading specialists focused on this family have included Borowski, Damoiseau, Lesne, Rai, Reichardt, and Vrydagh, all of which studied this family thoroughly. This updated catalogue is published approximately fifteen years after the first version of the previous catalogue (Borowski & Węgrzynowicz 2007a; modification of Borowski 2020; T. Borowski 2020, 2021). Since then, a number of new species and genera (including fossils) have been described, and there has also been a change in the approach to the higher taxonomy (Liu & Schönitzer, 2011) that is used here. Several problems of species previously classified as „incertae sedis" with an unclear family categorization have been resolved. On the other hand, some new issues emerged. Nevertheless, the key achievement is more accurate specification of individual species' distribution by countries in individual regions.

The bibliography includes all primary sources with taxon descriptions (including those which are currently classified in other families). With the faunistic sources, only a selection of sources is included – we have taken into consideration catalogues covering larger geographic areas, works covering specific regions or taxonomy groups with valuable faunistic facts. This data tended

TABLE 1 Number of genera in subfamilies in individual biogeographic regions

Subfamily	P	N	E	I	Nt	A
Dinoderinae	6	4	5	7	4	3
Dysidinae	1			1	1	
Endecatominae	1	1				
Euderiinae						1
Lyctinae	7	4	8	5	5	10
Polycaoninae	2	2		2	2	
Psoinae	3	1	1	4	2	
Bostrichinae	35	26	26	25	30	19
†Alitrepaninae				1		

TABLE 2 Number of species in subfamilies in individual biogeographic regions

Subfamily	P	N	E	I	Nt	A
Dinoderinae	29	23	13	25	13	10
Dysidinae	1			1	1	
Endecatominae	2	3				
Euderiinae						1
Lyctinae	31	23	22	22	27	27
Polycaoninae	5	11		4	18	
Psoinae	6	2	2	4	4	
Bostrichinae	132	89	171	77	113	45
†Alitrepaninae				11		

TABLE 3 Number of genera and species in subfamilies

Subfamily	Genera	Species
Dinoderinae	7	59
Dysidinae	2	2
Endecatominae	1	4
Euderiinae	1	1
Lyctinae	13	62
Polycaoninae	3	29
Psoinae	7	15
Bostrichinae	63	416
†Alitrepaninae	1	11
Total	**98**	**599**

to be different for the Palearctic area where we have based our work on findings by Borowski (2007b) and several national catalogues that are not included in the bibliography individually.

In most species, the development takes place in wood of trees and bushes – mainly in trunks, branches, and roots, but also in cut timber and various wooden products – and in various conditions in terms of wood humidity – from living and withering or withered wood to completely dry old wood. Some species live in symbiosis with fungi.

A total of 19 fossil species in 8 genera are known, with 4 genera being fossil only, with a total of 12 species.

Material and Methods

All the names, years of publications, page numbers and other data included in the catalogue have been carefully checked with the original literature. The names of subfamilies and genera are given as headings; after them, there are type genera (for subfamilies) and type species (for genera), as well as synonyms (also with their type genera and type species). The valid specific name is followed by the name of its author, the year of publication and the page on which it appeared.

The type depository of all the species if not revised, is given as published, or the respective collections in which they are held are quoted as in Horn et al. (1990a, b).

Citations of papers consist of the author's name and the year of publication. With more than one cited paper published in the same year by a given author, the first one is marked by the addition of the letter a, the second by the letter b etc.

An alphabetical list of synonyms, names of species transferred, nomina nuda, and invalid names, used in the combination with a given generic name or its synonym is stated for each genus after the list of the known species.

For some others, marked as „nomen nudum" the designation „homonym" was also used, although it is not completely useful. It was used for some species, if a certain name was used by one author in one work in one genus more than once, even on one page. The reason was an effort to draw attention to this, so that there would be no assumption of an error in our publication, such as an erroneous repeated citation.

In some cases, new homonyms were discovered, but in accordance with the principles of maintaining the stability of nomenclature with regard to their origin and the fact that they now belong to different species, in accordance with the rules of the ICZN, names are not established, only the indication that it is homonym was made.

If the pagination in the papers was indicated in Roman numerals, it was converted to Arabic numerals. Roman numerals were replaced by Arabic ones (years of journal, pagination in some journals and in book introductions or appendices).

Acronyms and marks used in the catalogue:

? unknown / uncertain
† fossil taxon
i introduced
HN homonym

IN infrasubspecific name
LC lapsus calami (error)
ND nomen dubium
NN nomen nudum
NO nomen oblitum
RN replacement name (nomen novum)
sic! error – actually written as indicated
var. variety

One of the key elements in this text is the geographical distribution of individual species. It includes data compiled from the bibliography completed with information based on findings from museum and private collections, including those of the authors. When working on this text, we had to deal with numerous problems. Firstly, there is not a clear conversion for the late 19th and early 20th century geography due to the colonial past. As colonies often broke up into several successor states (or alternatively, border areas were annexed to another state) and furthermore, original (indigenous) names were used sometimes, identification is not reliable in some cases. We did not use such misleading data, however, there might have been some misinterpretations although we made all effort to avoid mistakes. This situation is also connected with the emergence of new states (such as Sudan vs. South Sudan). It was not always possible to define the location exactly and unambiguously – in such indeterminate cases, the location was stated in the original state before the division. Vrydagh and some other authors use distribution maps in their publications, marking not only locations with specific ideas (which may not be stated in the text), but also showing potential distribution in the given region. This type of distribution was also used by Borowski & Węgrzynowicz (2007a).

In case of big territorial units (states), their division in smaller regions was used. In most cases, such division into smaller units was based on natural conditions, but occasionally, other principles were applied. In the Palearctic realm, the categorization for Russia and China was used according to Löbl & Smetana 2007. The categorization for North America is based on the series of beetle catalogues for this area, such as White 1982. The categorization of India is partly based on Löbl & Smetana 2007 for the Palearctic part, while the rest was categorized by us the authors according to the natural conditions (Butze 1959; Hendrych 1984). The same sources were used for Mexico and Brazil – the foundation comes from natural conditions aggregated and simplified as groups of individual states).

Mexico is divided into two regions – Baja California (North and South) and Sonora are classified in the Nearctic region (a number of catalogs assign northern Mexico to the Nearctic Region, which we used to), the rest of Mexico is then classified in the Neotropical Region with regard to climatic and vegetation conditions.

Whole area of China is classified as part of the Palaearctic Region. Argentina and Chile are divided geopolitically into the northern and southern parts. In this case, we have not used the work of Morrone (2001) who introduced a novel approach to the Neotropical Region – he did so on the supranational level, which is not applicable for the purposes of this catalogue. In other cases, he provides details, however, these are also not applicable for our purposes. Australia is divided – more or less – according to individual states. Indonesia is categorized based on individual islands or islands groups. In the case of Malaysia, the continental part and the Island of Borneo have been differentiated.

In the division of selected "big" states into smaller regions (except for Nearctic Region where data is rather detailed), the resulting list of regions with the given species' distribution may not be definitive. In some cases, more detailed data are not available in the bibliographical sources.

The basis of higher classification from the level of tribes is according to Bouchard et al. (2011) and Cai et al. (2022) with minor corrections subsequently made by some specialists for this family.

Note: Dejean's, Sturm's, Melsheimer's and some other names of new species in their catalogues are nomina nuda. These names consistently go against the principle of priority in nomenclature, preferring instead the most used name over the oldest. Many of the species listed here will be representatives of the family Curculionidae, subfamily Scolytinae and some other families. This also applies to the genus *Apate*, *Bostrichus*, *Ptinus*, *Lyctus* and others.

Distribution of Species: Acronyms and Maps

A: Australian Region

AS Australia
- ASE New South Wales and Victoria
- ASH Lord Howe I. and Norfolk I.
- ASN Northern Territory
- ASQ Queensland
- ASS Southern Australia
- AST Tasmania
- ASW Western Australia

BS Bismarck Is.
CIS Cook Is.
CO Micronesia (Caroline Is.)
FJ Fiji
FPS French Polynesia
- FPG Gambier I.
- FPM Marquesc Is.
- FPT Tuamotu Is.
- FPU Tubuai Is.

HI Hawaiian Is.
IA Indonesia
- IAN New Guinea
- SH Sumba, Flores, Timor, Buru, Seram, Halmahera, Bali

KC Kermadec Is.
KG Kerguelen Is.
KB Kiribati
MI Mariana Is.
MH Marshall Is.
NA New Caledonia Is.
NAU Nauru
NH Vanuatu (New Hebrides Is.)
NZ New Zealand
PI Palau
PW Papua – New Guinea
SJ Samoa
SS Solomon Is.
TA Tahiti

TI Tonga
TRC Micronesia (Truck I.)
TV Tuvalu

E: Afrotropical Region

AA Angola
BD Burundi
BF Burkina Faso
BN Benin
BW Botswana
CD Chad
CF Central African Republica
CG Congo (Brazzaville)
CK Cameroon
CM Comoro Is.
CV Cape Verde Is.
DJ Djibouti
EH Ethiopia
EQ Equatorial Guinea
ER Eritrea
GA Gambia
GH Ghana
GO Gabon
GS Guinea-Bissau
GX Guinea
IV Ivory Coast
KY Kenya
LH Lesotho
LI Liberia
MB Madagascar
MAY Mayotte I.
ML Mali
MU Mauritius I.
MW Malawi
MT Mauretania
MZ Mozambique
NB Namibia
NI Niger

NX Nigeria
RI Reunion I.
RW Rwanda
SF South Africa
SG Senegal
SO Somalia
SR Sierra Leone
SSU South Sudan
STI St. Thomas I. and Prince I.
SU Sudan
SVA St. Helen I.
SWA Swaziland
SYC Seychelles Is.
TG Togo
TZ Tanzania (incl. Zanzibar)
UG Uganda
ZA Zaire (Kinshasa)
ZB Zambia
ZI Zimbabwe

I: Indomalaian Region

AI Andaman Is. and Nicobar Is.
BG Bangladesh
BM Burma (Myanmar)
BX Brunei
CA Cambodia
CHI Christmas I.
CX Ceylon (Sri Lanka)
IA Indonesia
- IAB Borneo
- IAC Celebes (Sulawesi)
- IAJ Java, Lombok
- IAS Sumatra

ID India
- IDB Bihar [BI], Jharkhand [JH], Orissa [OR], West Bengal (without Darjeeling distr.) [WBS]
- IDC Gujarat [GU], Madhya Pradesh [MP], Maharashitra [MA]

- IDE Assam [AS], Manipur [MR], Meghalaya [ME], Mizoram [MZ], Nagaland [NA], Tripura [TR]
- IDH Andhra Pradesh [AN], Chhattisgarh [CH], Telangány [TE]
- IDP Haryana [HA], Punjab [PU], Rajasthan [RA]
- IDS Tamil Nadu [TN]
- IDW Goa [GO], Karnataka [KN], Kerala [KE]

LO Laos
MY Malaysia
- MYC continental Malaysia
- MYS Sarawak and Sabah (Borneo)

MS Maldives Is. + Laccadive Is.
SIG Singapore
PH Philippines
TH Thailand
VT Vietnam

N: Nearctic Region

CN Canada
- CNA Arctic Territory (Northwest Territories [NT], Yukon Territory [YT])
- CNL Laurentian Territory (Ontario [ON], Quebec [PQ])
- CNM Maritime Territory (New Brunswick [NB], Newfoundland [NF], Nova Scotia [NS], Prince Edward I. [PE], St. Piere-Miquelon [PM])
- CNN Pacific Northwest Territory (British Columbia [BC])
- CNP Prairie Territory (Alberta [AB], Saskatchewan [SK], Manitoba [MB])

DE DEA Greenland I. [GL]
MX Mexico
- MXC Baja California Norte (BCN), Baja California Sur (BCS), Sonora (SON)

US United States of America
- USA Arctic Territory (Alaska [AK])
- USC Californian Territory (California [CA], Nevada [NV])
- USD Middle States Territory (Iowa [IA], Illinois [IL], Indiana [IN], Kansas [KS], Kentucky [KY], Missouri [MO], Nebraska [NE], Ohio [OH])
- USE Southeast Territory (Alabama [AL], Arkansas [AR], Florida [FL], Georgia [GA], Louisiana [LA], Mississippi [MS], North Carolina [NC], South Carolina [SC], Tennessee [TN])
- USG New England Territory (Connecticut [CT], Massachusetts [MA], Maine [ME], New Hampshire [NH], Rhode Island [RI], Vermont [VT])
- USL Laurentian Territory (Michigan [MI], Minnesota [MN], Wisconsin [WI])

USN Pacific Northwest Territory (Idaho [ID], Oregon [OR], Washington [WA])

USO Southwest Territory (Arizona [AZ], New Mexico [NM], Oklahoma [OK], Texas [TX])

USP Prairie Territory (Montana [MT], North Dakota [ND], South Dakota [SD])

USU Mountain Territory (Colorado [CO], Utah [UT], Wyoming [WY])

UST Middle Atlantic Territory (District of Columbia [DC], Delaware [DE], Maryland, [MD], New Jersey [NJ], New York [NY], Pennsylvania [PA], Virginia [VA], West Virginia [WV])

Nt: Neotropical Region

AT Antigua and Barbuda Is.

AY Argentina

AYN Argentina north (Catamarca [CM], Cordoba [CD], Corrientes [CO], Entre Rios [ER], Formosa [FO], Chaco [CA], Jujuy [JU], La Rioja [RI], Mendosa [ME], Misiones [MI], Salta [SA], San Juan [SH], San Luis [SL], Santa Fe [SF], Santiago del Estero [SE], Tucaman [TU])

AYS Argentina south (Buenos Aires [BA], Chubut CH]; La Pampa [PA], Neuquén [NE], Rio Negro [RN], Santa Cruz [SC], Tiera del Fuego]TF])

FK Falkland Is.

BI Bahamas Is.

BL Belize

BR Barbados I.

BV Bolivia

BZ Brazil

BZA Acre [AC], Amapá [AP], Amazonas [AM], Pará [PA], Roraima [RO]

BZC Esperito Santo [ES], Goiás [GO], Minas Gerais [MI], Municipo Neutro [MN], Rio de Janeiro [RJ], Tocantinis [TO]

BZE Alagoas [AL], Bahia [BA], Ceará [CE], Maranhão [MA], Pãraiba [PB], Pernambuco [PE], Piauí [PI], Rio Grande de Norte [RGN], Sergipe [SE]

BZM Mato Grosso [MG], Mato Grosso do Sud [MGS], Rondônia [RD]

BZS Paraná [PR], Rio Grande do Sul [RG], Santa Catarina [SC], São Paulo [SP]

CB Colombia

CC Costa Rica

CL Chile

CLN Chile north (Antofagasta [AN], Arica y Parinacota [AP], Atacama [AT], Bío Bío [BB], Coquimbo [CQ], La Aracuania [AR], Libertador

Bernardo O'Higgins [BH], Los Ríos [LR], Maule [MA], Metropolitana de Santiago [SA], Ñuble [NU], Tarapacá [TA], Valparaiso [VA])

CLS Chile south (Aysén del General Ibañez del Campo [AY]; Los Lagos [LL], Magallanes y Antártica Chilena [MA])

CMS Cayman Is.

CU Cuba

DO Dominica

DR Dominican Republic

EC Ecuador

ES El Salvador

FG French Guiana

GI Galapagos Is.

GL Guadeloupe I.

GN Grenada I.

GT Guatemala

GU Guyana

HA Haiti I.

HO Honduras

JC Jamaica

LW Leeward Is. (Dutch Antilles)

MNT Montserrat I.

MQ Martinique I.

MX Mexico

MXE Coahuila (CO), Nuevo León (NL), San Luis Potosí (SLP), Tamaulipas (TA)

MXM Distrito Federal (DF), Guanajuato (GU), Hidalgo (HD), Mexico (MX), Michoacán (MI), Querétaro (QU), Tlaxcala (TL)

MXS Aguascalientes (AG), Chihuahua (CH), Durango (DU), Zacatecas (ZA)

MXT Chiapas (CI), Guerrero (GR), Morelos (MO), Oaxaca (OA), Puebla (PB), Tabasco (TB), Veracruz (VC)

MXW Colima (CL), Jalisco (JA), Nayarit (NA), Sinaloa (SI)

MXY Campeche (CA), Quintana Roo (QR), Yucatan (YU)

NG Nicaragua

PA Panama

PE Peru

PG Paraguay

PR Puerto Rico I.

SM Suriname

SN St. Lucia I.

SQ St. Christopher I. and Nevis Is.

STB St. Barthelemy I.

STV St. Vincent I. and Grenadine Is.
STT St. Thomas I.
TT Trinidad and Tobago Is.
UR Uruguay
VE Venezuela
VI Virgin Is.

P: Palaearctic Region

AB Azerbaijan
AE United Arab Emirates
AF Afghanistan
AG Algeria
AL Albania
AN Andora
AR Armenia
AZ Azores I.
AU Austria
BA Bahrain
BE Belgium
BH Bosnia and Herzegovina
BT Bhutan
BU Bulgaria
BY Byelorussia
CH China
- CE Central Territory (Anhui [ANH], Hubei [HUB], Hunan [HUN], Jiangxi [JIX], Jiangsu [JIA], Shanghai [SHG], Zhejiang [ZHE])
- HAI Hainan I.
- NE Northeast Territory (Heilongjiang [HEI], Jilin [JIL], Liaoning [LIA])
- NO Northern Territory (Beijing [BEI], Hebei [HEB], Henan [HEN], Nei Mongol [NMO], Ningxia [NIN], Shaanxi [SHA], Shanxi [SHX], Shandong [SHN], Tianjin [TIA])
- NW Northwest Territory (Xinjiang [XIN], Gansu [GAN])
- SE Southeastern Territory (Fujian [FUJ], Guangdong [GUA], Guangxi [GUX], Hongkong [HKG], Macao [MAC])
- SW Southwestern Territory (Chongquing [CHQ], Guizhou [GUI], Sichuan [SCH], Yunnan [YUN])
- TAI Taiwan I.
- WP Western Plateau (Xizang [XIZ], Qinghai [QIN])

CR Croatia

CI Canary Is. (incl. Selvages Is.)
CY Cyprus
CZ Czechia
DE Denmark
EG Egypt (incl. Sinai)
EN Estonia
FA Faeroe Is.
FI Finland
FR France (incl. Corsica, Monaco)
GB Great Britain (incl. Channel Is.)
GE Germany
GG Georgia
GR Greece (incl. Crete and other islands)
HU Hungary
IC Iceland
ID India
- AP Arunachal Pradesh [AP]
- HP Himachal Pradesh [HP]
- KA Jammu [JA) Kashmir [KA]
- SD Sikkim [SK], Darjeeling distr. [WBN]
- UP Uttaranchal [UT], Uttar Pradesh [UP]

IN Iran
IQ Iraq
IR Ireland
IS Israel
IT Italy (incl. Sardinia, Sicily and San Marino)
JA Japan
JO Jordan
KU Kuwait
KY Kyrgyzstan
KZ Kazakhstan
LA Latvia
LB Libya
LE Lebanon
LS Liechtenstein
LT Lithuania
LU Luxembourg
MC North Macedonia
MA Malta
MD Moldavia

ME Montenegro
MG Mongolia
MO Morocco (incl. Western Sahara)
MR Madeira Is.
NC North Korea
NL The Netherlands
NP Nepal
NR Norway
OM Oman
PA Pakistan
PL Poland
PT Portugal
QA Qatar
RO Romania
RU Russia
 CT Central European Territory
 ES East Siberia
 FE Far East (incl. Sakhalin I.)
 NT North European Territory
 ST South European Territory
 WS West Siberia
SA Saudi Arabia
SB Serbia and Kosovo
SC South Korea
SK Slovakia
SL Slovenia
SP Spain (incl. Gibraltar and Baleares Is.)
SV Sweden
SZ Switzerland
SY Syria
TD Tadzhikistan
TM Turkmenistan
TR Turkey
TU Tunisia
UK Ukraine
UZ Uzbekistan
YE Yemen
 YEC – Yemen continental
 YES – Yemen – Socotra Is.
YU former Yugoslavia (without specific country)

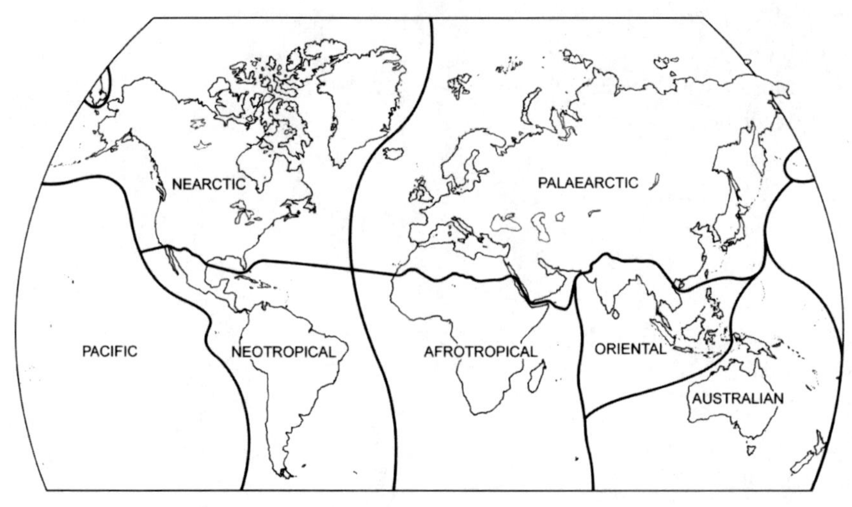

MAP 1 The limits of the geographic Regions as defined for the purpose of this catalogue (according to Löbl & Smetana 2007)

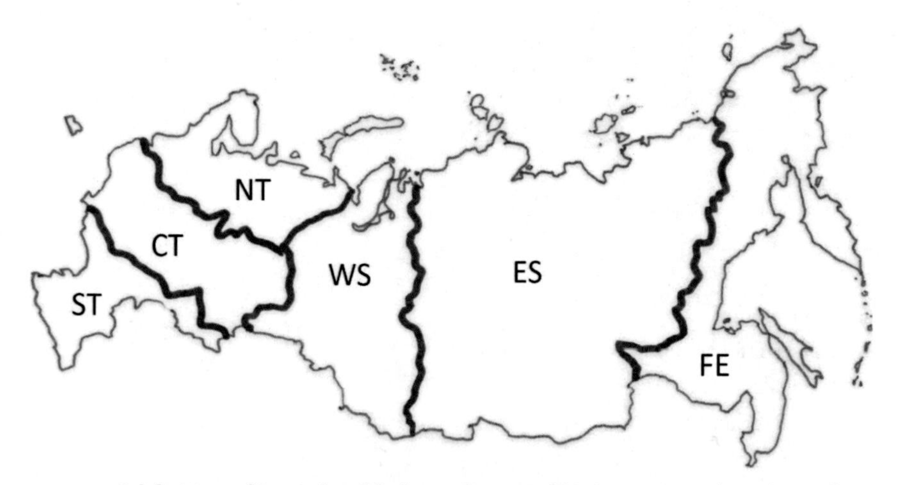

MAP 2 Subdivisions of Russia (modified according to Löbl & Smetana 2007): **CT** (Central European Territory); **ES** (East Siberia); **FE** (Far East [incl. Sakhalin I.]); **NT** (North European Territory); **ST** (South European Territory); **WS** (West Siberia)

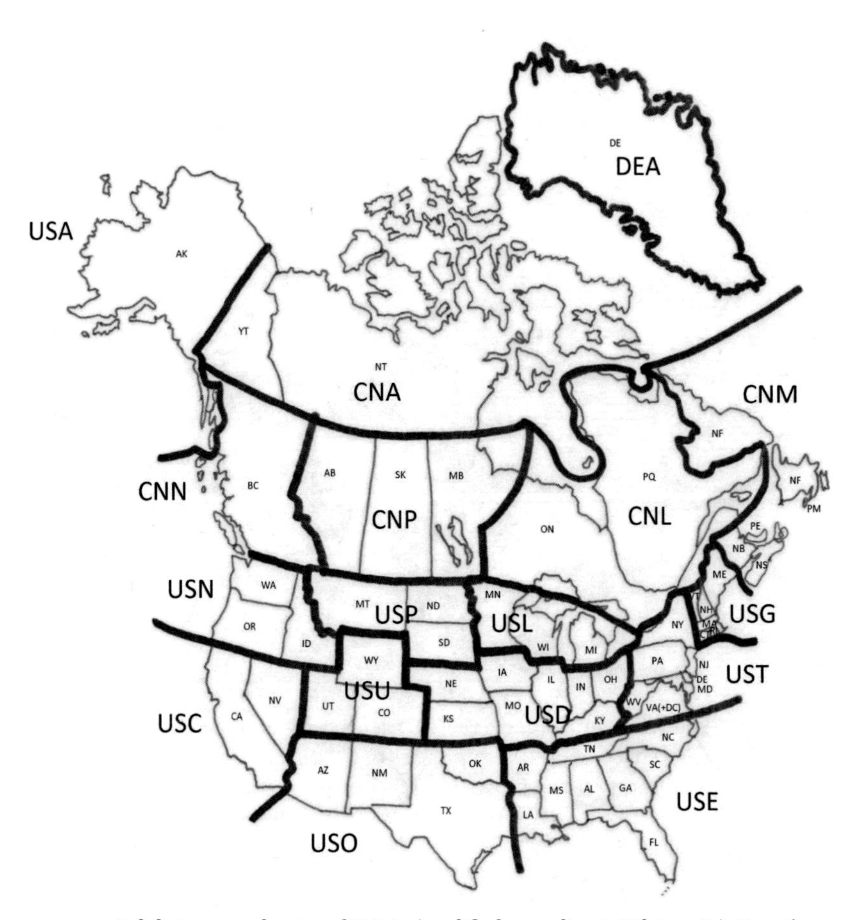

MAP 3 Subdivisions and states of U.S.A. (modified according to White 1982): *Canada*: **CNA** – Arctic Territory (Northwest Territories [NT], Yukon Territory [YT]); **CNL** – Laurentian Territory (Ontario [ON], Quebec [PQ]); **CNM** – Maritime Territory (New Brunswick [NB], Newfoundland [NF], Nova Scotia [NS], Prince Edward I. [PE], St. Piere-Miquelon [PM]); **CNN** – Pacific Northwest Territory (British Columbia [BC]); **CNP** – Prairie Territory (Alberta [AB], Saskatchewan [SK], Manitoba [MB]). *Denmark*: **DEA** (Greenland I.). *United States of America*: **USA** – Arctic Territory (Alaska [AK]); **USC** – Californian Territory (California [CA], Nevada [NV]); **USD** – Middle States Territory (Iowa [IA], Illinois [IL], Indiana [IN], Kansas [KS], Kentucky [KY], Missouri [MO], Nebraska [NE], Ohio[OH]); **USE** – Southeast Territory (Alabama [AL], Arkansas [AR], Florida [FL], Georgia [GA], Louisiana [LA], Mississippi [MS], North Carolina [NC], South Carolina [SC], Tennessee [TN]); **USG** – New England Territory (Connecticut [CT], Massachusetts [MA], Maine [ME], New Hampshire [NH], Rhode Island [RI], Vermont [VT]); **USL** – Laurentian Territory (Michigan [MI], Minnesota [MN], Wisconsin [WI]); **USN** – Pacific Northwest Territory (Idaho [ID], Oregon [OR], Washington [WA]); **USO** – Southwest Territory (Arizona [AZ], New Mexico [NM], Oklahoma [OK], Texas [TX]); **USP** – Prairie Territory (Montana [MT], North Dakota [ND], South Dakota [SD]); **USU** – Mountain Territory (Colorado [CO], Utah [UT], Wyoming [WY]); **UST** – Middle Atlantic Territory (District of Columbia [DC], Delaware [DE], Maryland, [MD], New Jersey [NJ], New York [NY], Pennsylvania [PA], Virginia [VA], West Virginia [WV])

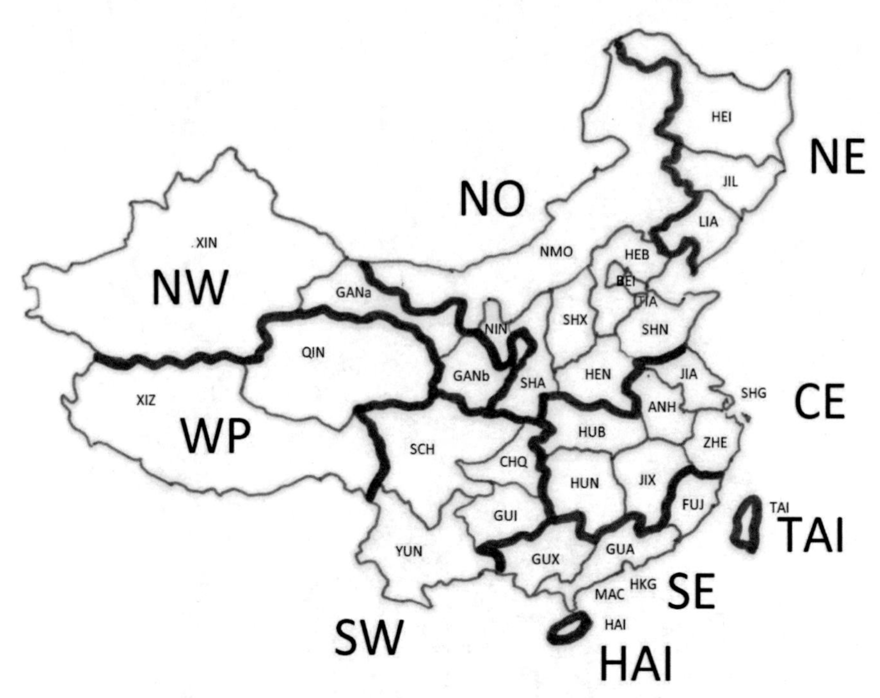

MAP 4 Subdivisions and provinces of the People's Republic of China (modified according to Löbl & Smetana 2007): **CE** – Central Territory (Anhui [ANH], Hubei [HUB], Hunan [HUN], Jiangxi [JIX], Jiangsu [JIA], Shanghai [SHG], Zhejiang [ZHE]); **HAI** – Hainan I. ([HAI]); **NE** – Northeast Territory (Heilongjiang [HEI], Jilin [JIL], Liaoning [LIA]); **NO** – Northern Territory (Beijing [BEI], Hebei [HEB], Henan [HEN], Nei Mongol [NMO], Ningxia [NIN], Shaanxi [SHA], Shanxi [SHX], Shandong [SHN], Tianjin [TIA]); **NW** – Northwest Territory (Xinjiang [XIN], Gansu [GAN]); **SE** – Southe/astern Territory (Fujian [FUJ], Guangdong [GUA], Guangxi [GUX], Hongkong [HKG], Macao [MAC]); **SW** – Southwestern Territory (Chongquing [CHQ], Guizhou [GUI], Sichuan [SCH], Yunnan [YUN]); **TAI** – Taiwan ([TAI]); **WP** – Western Plateau (Xizang [XIZ], Qinghai [QIN])

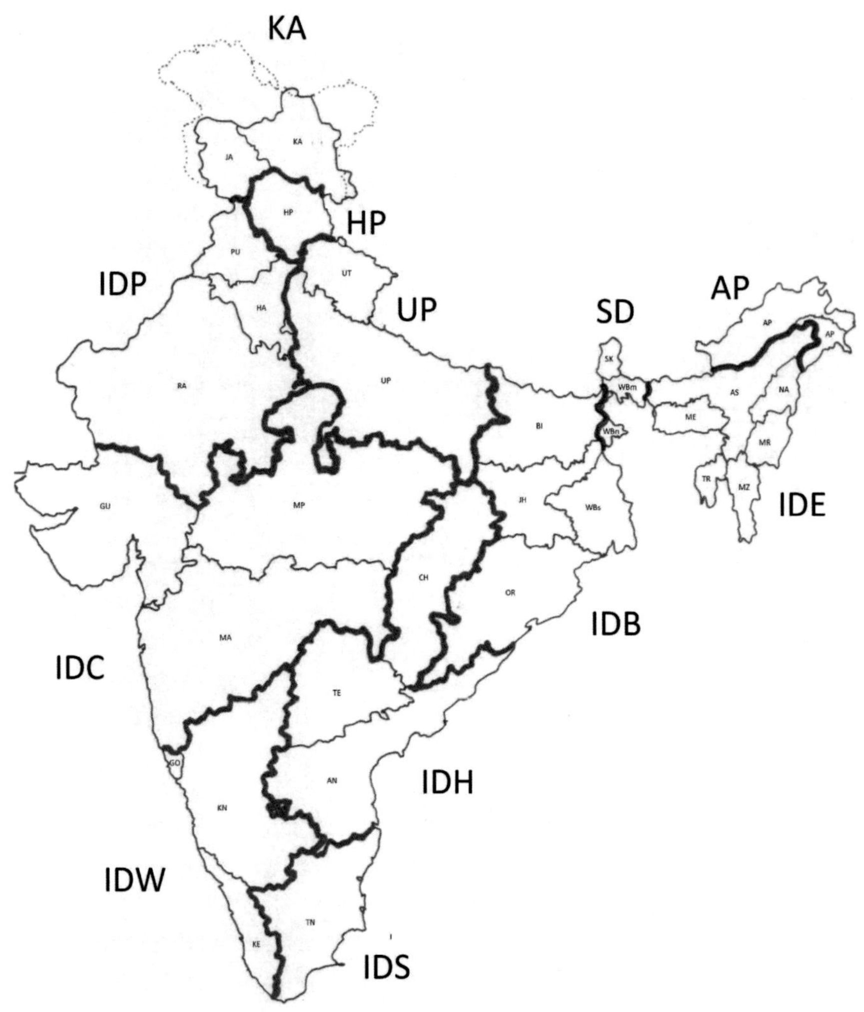

MAP 5 Subdivisions and states of India: *Palaearctic Region* – **AP** (Arunachal Pradesh [AP]); **HP** (Himachal Pradesh [HP]); **KA** (Jammu [JA], Kashmir [KS]); **SD** (Sikkim [SK] and West Bengal-Darjeeling distr. [WBN]); **UP** (Uttaranchal [UT], Uttar Pradesh [UP]). *Oriental Region* – **IDB** (Bihar [BI], Jharkhand [JH], Orissa [OR], West Bengal (without Darjeling distr.) [WBS]); **IDC** (Gujarat [GU], Madhya Pradesh [MP], Maharashitra [MA]); **IDE** (Assam [AS], Manipur [MR], Meghalaya [ME], Mizoram [MZ], Nagaland [NA], Tripura [TR]); **IDH** (Andhra Pradesh [AN], Chhattisgarh [CH], Telangány [TE]); **IDP** (Haryana [HA], Punjab [PU], Rajasthan [RA]); **IDS** (Tamil Nadu [TN]); **IDW** (Goa [GO], Karnataka [KN], Kerala [KE])

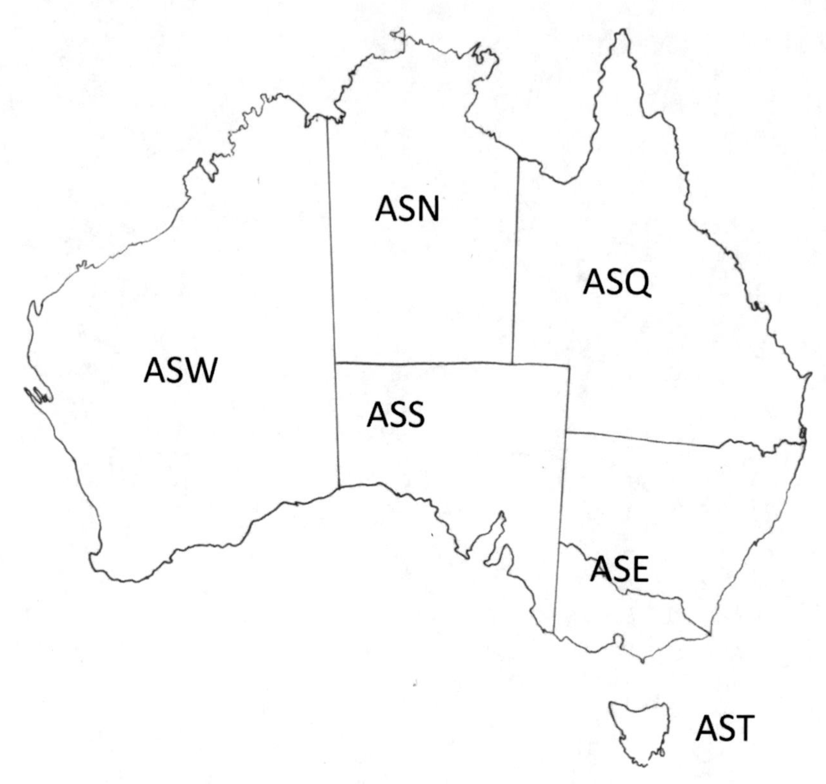

MAP 6 Subdivisions and states of Australia: **ASE** (New South Wales and Victoria); **ASN** (Northern Territory); **ASQ** (Queensland); **ASS** (Southern Australia); **AST** (Tasmania); **ASW** (Western Australia)

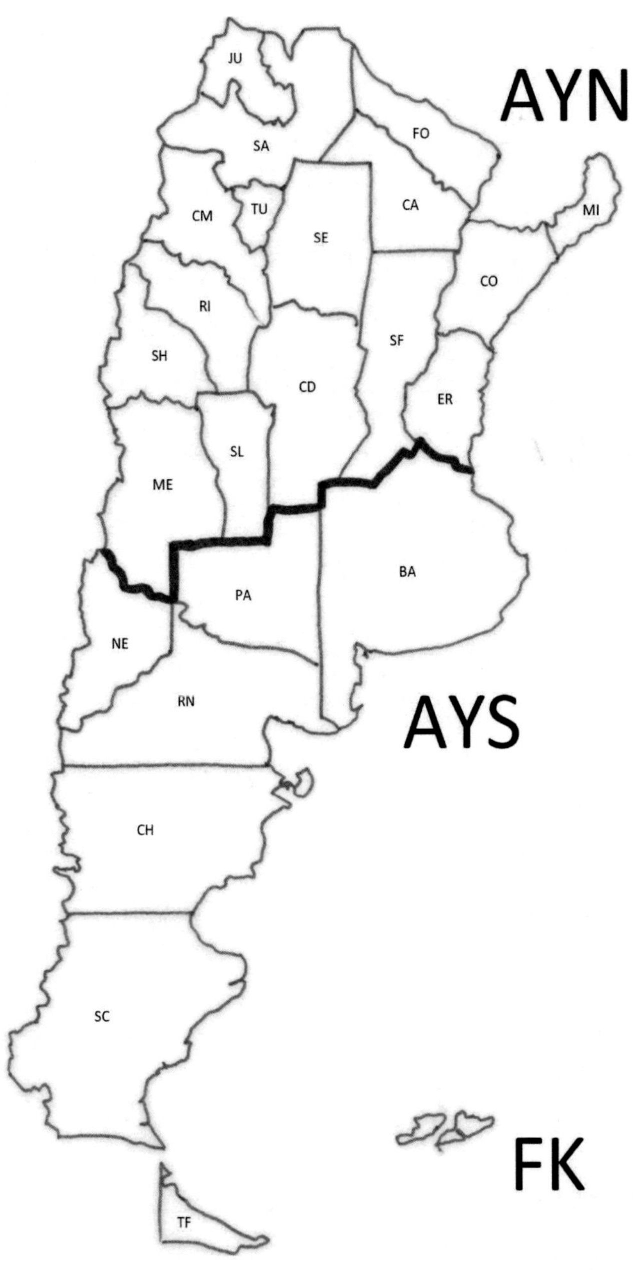

MAP 7 Subdivisions and states of Argentina: **AYN** – Argentina North (Catamarca [CM], Cordoba [CD], Corrientes [CO], Entre Rios [ER], Formosa [FO] , Chaco [CA], Jujuy [JU], La Rioja [RI], Mendosa [ME], Misiones [MI], Salta [SA], San Juan [SH], San Luis [SL], Santa Fe [SF] , Santiago del Estero [SE], Tucumán [TU]); **AYS** – Argentina South (Buenos Aires [BA], Chubut [CH]; La Pampa [PA],Neuquén [NE], Rio Negro [RN], Santa Cruz [SC], Tierra del Fuego [TF]); **FK** – Falkland Is.

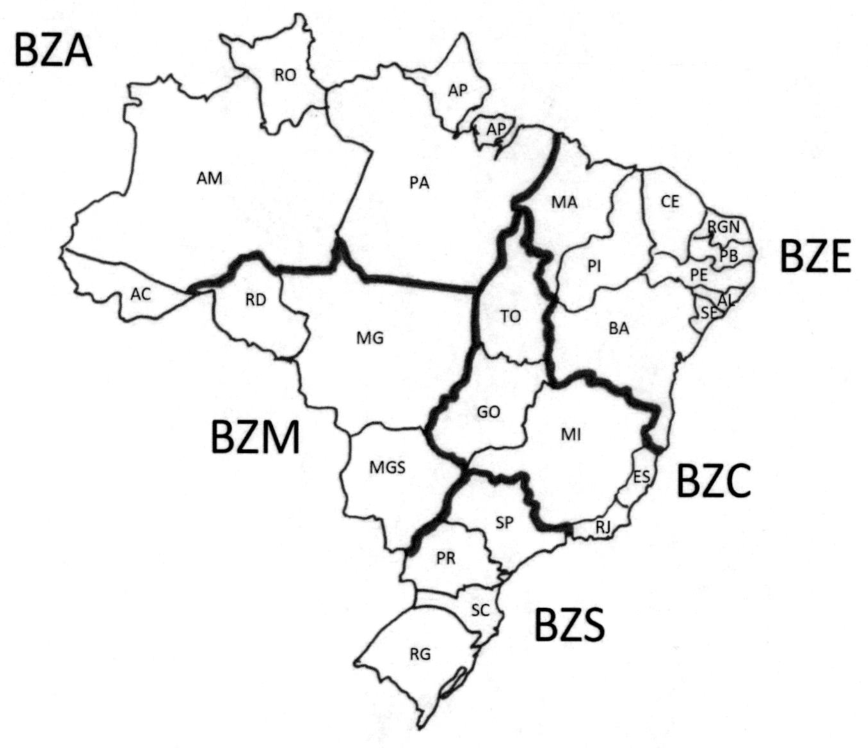

MAP 8 Subdivisions and states of Brazil: **BZA** (Acre [AC], Amapá [AP], Amazonas [AM], Pará [PA], Roraima [RO]); **BZC** (Esperito Santo [ES], Goiás [GO], Minais Gerais [MI], Rio de Janeiro [RJ], Tocantinis [TO]); **BZE** (Alagoas [AL], Bahia [BA], Ceará [CE], Maranhão [MA], Pãraiba [PB], Pernambuco [PE], Piauí [PI], Rio Grande de Norte [RGN], Sergipe [SE]); **BZM** (Mato Grosso [MG], Mato Grosso do Sul [MGS], Rondônia [RD]); **BZS** (Paraná [PR], Rio Grande do Sul [RG], Santa Catarina [SC], São Paulo [SP])

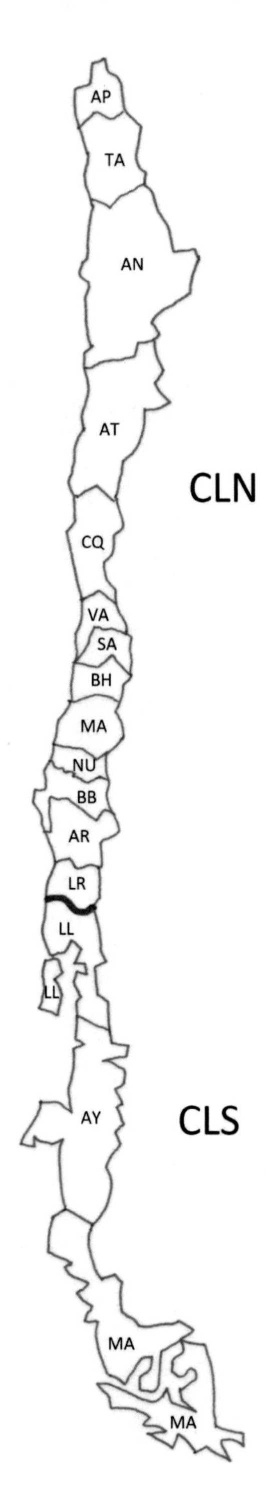

MAP 9
Subdivisions and states of Chile: **CLN** – Chile north (Antofagasta [AN], Arica y Parinacota [AP], Atacama [AT], Bío Bío [BB], Coquimbo [CQ], La Aracuania [AR], Libertador Bernardo O'Higgins [BH], Los Ríos [LR], Maule [MA], Metropolitana de Santiago [SA], Ñuble [NU], Tarapacá [TA], Valparaiso [VA]); **CLS** – Chile south (Aysén del General Ibañez del Campo [AY]; Los Lagos [LL], Magallanes y Antártica Chilena [MA])

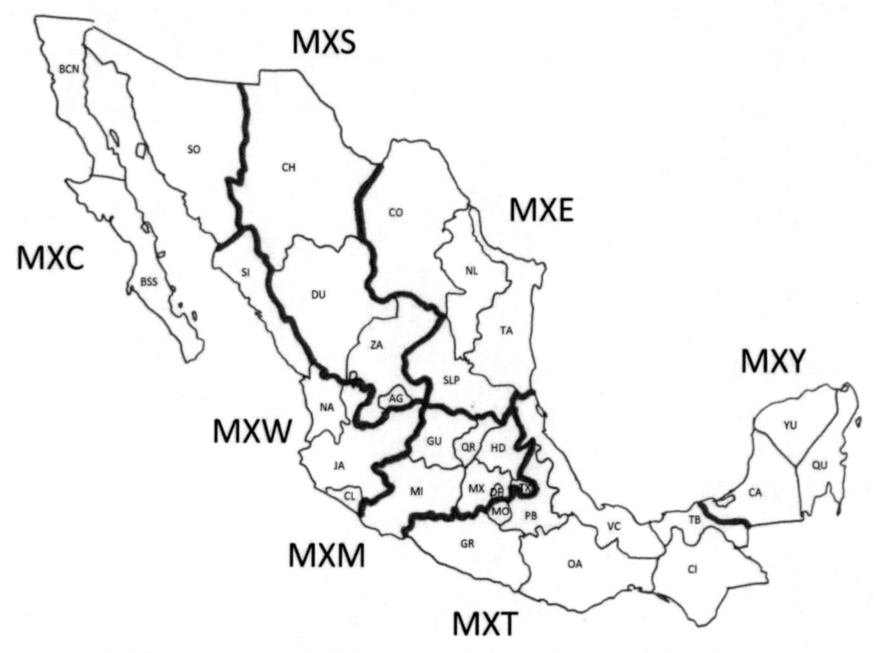

MAP 10 Subdivisions and states of Mexico: **MXC** (Baja California Norte [BCN], Baja California Sur [BCS], Sonora [SO]); **MXE** (Cohauila [CO], Nuevo León [NL], San Luis Potosí [SLP], Tamaulipas [TA]); **MXM** (District Federal [DF], Guanajuato [GU], Hidalgo [HD], Mexico [MX], Michoacán [MI], Querétaro [QR], Tlaxcala [TX]); **MXS** (Aguascalientes [AG], Chihuahua [CH], Durango [DU], Zacatecas [ZA]); **MXT** (Chiapas [CI], Guerrero [GR], Morelos [MO], Oaxaca [OA], Puebla [PB], Tabasco [TB], Veracruz [VC]); **MXW** (Colima [CL], Jalisco [JA], Nayarit[NA], Sinaloa [SI]); **MXY** (Campeche [CA], Quintana Roo [QU], Yucatan [YU])

Systematics of Bostrichidae

According to Borowski & Węgrzynowicz, 2007a	According to Liu & Schönitzer, 2011
1. Lyctinae	1. Lyctinae
1.1. Lyctini	1.1. Cephalotomini
1.1.1. Acantholyctus	1.1.1. Cephalotoma
1.1.2. Lycthoplites	1.2. Lyctini
1.1.3. Lyctodon	1.2.1. Acantholyctus
1.1.4. Lyctoxylon	1.2.2. Loranthophila[b]
1.1.5. Lyctus	1.2.3. Lycthoplites
1.1.6. Minthea	1.2.4. Lyctodon
1.2. Trogoxylini	1.2.5. Lyctoxylon
1.2.1. Cephalotoma	1.2.6. Lyctus
1.2.2. Lyctoderma	1.2.7. Minthea
1.2.3. Lyctopsis	1.3. Trogoxylini[a]
1.2.4. Phyllyctus	1.3.1. Lyctopsis
1.2.5. Tristaria	1.3.2. Phyllyctus
1.2.6. Trogoxylon	1.3.3. Tristaria
2. Psoinae	1.3.4. Trogoxylon
2.1. Chileniini	1.3.5. Trogoxylyctus[b]
2.1.1. Chilenius	2. Dysidinae
2.2. Psoini	2.1. Apoleon
2.2.1. Coccographis	2.2. Dysides
2.2.2. Heteropsoa	3. Euderinae
2.2.3. Psoa	3.1. Euderia
2.2.4. Psoidia	4. Endecatominae
2.2.5. Stenomera	4.1. Endecatomus
3. Polycaoninae	5. Dinoderinae
3.1.1. Melalgus	5.1.1. Dinoderopsis[a]
3.1.2. Polycaon	5.1.2. Dinoderus
4. Euderinae	5.1.2.1. Dinoderus (s. str.) need to change margins
4.1.1. Euderia	5.1.2.2. Dinoderastes

a Using only newly (removed) by Borowski & Węgrzynowicz (2012)
b Newly described species or newly status, after Liu & Schönitzer (2011)

(*cont.*)

According to Borowski & Węgrzynowicz, 2007a	According to Liu & Schönitzer, 2011
5. Dysidinae	5.1.3. Orientoderus[a,b]
5.1.1. Apoleon	5.1.4. Prostephanus
5.1.2. Dysides	5.1.5. Rhizoperthodes
6. Endecatominae	5.1.6. Rhyzopertha
6.1.1. Endecatomus	5.1.7. Stephanopachys
7. Dinoderinae	6. Bostrichinae
7.1.1. Dinoderopsis	6.1. Apatini
7.1.2. Dinoderus	6.1.1. Apate
7.1.2.1. Dinoderastes	6.1.2. Dinapate
7.1.2.2. Dinoderus	6.1.3. Phonapate
7.1.3. Prostephanus	6.1.4. Protapate
7.1.4. Prostephanus	6.1.5. Xylomedes
7.1.5. Rhizopertodes	6.2. Bostrichini
7.1.6. Rhyzopertha	6.2.1. Amphicerus
7.1.7. Stephanopachys	6.2.2. Apatides
8. Bostrichinae	6.2.3. Bostrichus
8.1. Bostrichini	6.2.4. Bostrycharis
8.1.1. Amphicerus	6.2.5. Bostrichoplectron[b] **syn. nov.**
8.1.1.1. Amphicerus	6.2.6. Bostrichoplites
8.1.1.2. Caenophrada	6.2.7. Bostrychopsis
8.1.2. Apatides	6.2.8. Calophorus
8.1.3. Bostrichus	6.2.9. Dexicrates
8.1.4. Bostrycharis	6.2.10. Dolichobostrychus
8.1.5. Bostrychoplites	6.2.11. Discoclavata[c]
8.1.6. Calophorus	6.2.12. Heterobostrychus
8.1.7. Dexicrates	6.2.13. Lichenophanes
8.1.8. Dolichobostrychus	6.2.14. Micrapate
8.1.9. Dominikia	6.2.15. Neoterius
8.1.10. Heterobostrychus	6.2.16. Parabostrychus
8.1.11. Lichenophanes	6.2.17. Sinoxylodes
8.1.12. Megabostrichus	6.3. Sinoxylini
8.1.13. Micrapate	6.3.1. Calodectes

c Only fossil genera

(*cont.*)

According to Borowski & Węgrzynowicz, 2007a		According to Liu & Schönitzer, 2011	
8.1.14.	Neoterius	6.3.2.	Calodrypta
8.1.15.	Parabostrychus	6.3.3.	Calopertha
8.1.16.	Sinoxylodes	6.3.4.	Sinocalon
8.2.	Sinoxylini	6.3.5.	Sinoxylon
8.2.1.	Calodectes	6.3.6.	Xyloperthodes
8.2.2.	Calodrypta	6.4.	Xyloperthini
8.2.3.	Calopertha	6.4.1.	Amintinus
8.2.4.	Sinocalon	6.4.2.	Calonistes
8.2.5.	Sinoxylon	6.4.3.	Ctenobostrychus
8.2.6.	Xyloperthodes	6.4.4.	Dendrobiella
8.3.	Xyloperthini	6.4.5.	Enneadesmus
8.3.1.	Amintinus	6.4.6.	Gracilenta[b]
8.3.2.	Calonistes	6.4.7.	Infrantenna[b]
8.3.3.	Calophagus	6.4.8.	Mesoxylion
8.3.4.	Ctenobostrychus	6.4.9.	Octodesmus
8.3.5.	Dendrobiella	6.4.10.	Octomeristes
8.3.6.	Enneadesmus	6.4.11.	Paraxylion
		6.4.12.	Plesioxylion[b]
8.3.7.	Mesoxylion	6.4.13.	Plioxylion
8.3.8.	Octodesmus	6.4.14.	Psicula
8.3.9.	Paraxylion	6.4.15.	Scobicia
8.3.10.	Paraxylogenes	6.4.16.	Sifidius
8.3.11.	Plioxylion	6.4.17.	Tetrapriocera
8.3.12.	Psicula	6.4.18.	Xylion
8.3.13.	Scobicia	6.4.19.	Xylionopsis
8.3.14.	Sifidius	6.4.20.	Xylionulus
8.3.15.	Tetrapriocera	6.4.21.	Xylobiops
8.3.16.	Xylion	6.4.22.	Xyloblaptus
8.3.17.	Xylionopsis	6.4.23.	Xylobosca
8.3.18.	Xylionulus	6.4.24.	Xylocis
8.3.19.	Xylobiops	6.4.25.	Xylodectes
8.3.20.	Xyloblaptus	6.4.26.	Xylodeleis
8.3.21.	Xylobosca	6.4.27.	Xylodrypta
8.3.22.	Xylocis	6.4.28.	Xylogenes
8.3.23.	Xylodecta	6.4.29.	Xylomeira

(*cont.*)

According to Borowski & Węgrzynowicz, 2007a	According to Liu & Schönitzer, 2011
8.3.24. Xylodeiles	6.4.30. Xylopertha
8.3.25. Xylodrypta	6.4.31. Xyloperthella
8.3.26. Xylogenes	6.4.32. Xylophorus
8.3.27. Xylomeira	6.4.33. Xyloprista
8.3.28. Xylopertha	6.4.34. Xylopsocus
8.3.29. Xyloperthella	6.4.35. Xylothrips
8.3.30. Xylophorus	6.4.36. Xylotillus
8.3.31. Xyloprista	7. Psoinae
8.3.32. Xylopsocus	7.1. Chileniini
8.3.33. Xylothrips	7.1.1. Chilenius
8.3.34. Xylotillus	7.2. Psoini
9. Apatinae	7.2.1. Coccographis
9.1. Apatini	7.2.2. Heteropsoa
9.1.1. Apate	7.2.3. Psoa
9.1.2. Phonapate	7.2.4. Psoidia
9.1.3. Xylomedes	7.2.5. Sawianus[b]
9.2. Dynapatini	7.2.6. Stenomera
9.2.1. Dinapate	8. Polycaoninae
	8.1. Cretalgus[b,c]
	8.2. Melalgus
	8.3. Polycaon
	9. Alitrepaninae[b,c]
	9.1. Poinarinius[b,c]

Collections

AMSA	Australian Museum, Sydney, Australia
ANIC	Australian National Insect Collection, Division of Entomology, CSIRO, Canberra, Australia
BMNH	British Museum of Natural History [newly The Natural History Museum], London, United Kingdom
CACM	A. J. Cook Arthropod Research Collection, Michigan State University, USA
CASC	Department of Entomology, California Academy of Sciences, San Francisco, USA
CCHH	private collection of Christel and Hans Werner Hoffeins (Hamburg, Germany), later the specimens will be deposited at the Senckenberg Deutsches Entomologisches Institut (Müncheberg, Germany)
DEES	Universidade de São Paulo, Piracicaba, Brazil
DEFS	Universidade de São Paulo, São Paulo, Brazil
DEIC	Deutsche Entomologische Institut, Eberswalde, Germany
DFPC	Department of Forest Protection, Warsaw University of Life Sciences, Warsaw, Poland
EIHU	Entomological Institute, Faculty of Agriculture, Hokkaido University, Sapporo, Japan
FMNH	Field Museum of Natural History, Chicago, USA
GMUR	Geological Department and Museum University Rennes, France
HNHM	Zoological Department, Hungarian Natural History Museum, Budapest, Hungary
HUYC	Honghe University, Mengzi, Yunnan, China
IBSP	Instituto Biologico, São Paulo, Brazil
ICCM	Section of Insects and Spiders, Carnegie Museum of Natural History, Pittsburgh, USA
IFRI	Indian Forest Research Institute, Dehra Dun, Uttar Pradesh, India
ININ	Instituto Nacional de Investigación de las Ciencia Naturales, Buenos Aires, Argentina
IPPB	Institute of the Plant Pathology, Bogor, Indonesia
ISEA	Institute of Systematics and Ecology of Animals, Novosibirsk, Russia
ISNB	Institut Royal des Sciences Naturelles de Belgique, Brussels, Belgium
IZAS	Insect Collection, Institute of Zoology, Beijing, China
JHAC	Private Entomological Laboratory & Collection, Jiří Háva, Únětice u Prahy, Czech Republic
KUFA	Entomological Laboratory, Faculty of Agriculture, Kyushu University, Fukuoka, Japan

MACN Museo Argentina de Ciencias Naturales "Bernardino Rivadavia," Buenos Aires, Argentina
MAMU MacLeay Museum, University of Sydney, New South Wales, Australia
MCSN Museo Civico de Storia Naturale "Giacomo Doria" Genova, Italy
MCZC Entomology Department, Museum of Comparative Zoology, Harvard University, Cambridge, USA
MCZR Museo Civico di Zoologia, Rome, Italy
MGDL Museum d'Histoire Naturalle du Grand-Duchy de Luxembourg, Luxembourg
MLPA Universidad Nacional de La Plata, Museo de la Plata, La Plata, Argentina
MMMA Melbourne Museum, Melbourne, Australia
MNHN Museum National d'Histoire Naturelle, Paris, France
MNMS Museo Nacional de Ciencias Naturales, Madrid, Spain
MRAC Musée Royal de l'Afrique Centrale, Tervuren, Belgium
MSNM Museo Civico di Storia Naturale, Milano, Italy
MZPAN Muzeum i Institut Zoologii Polskiej Akademii Nauk, Warsa, Poland
MZSP Museu de Zoologia de Universade de São Paulo, São Paulo, Brazil
NHMB Entomology Department, Naturhistorischen Museum, Basel, Switzerland
NHMW Naturhistorisches Museum Wien, Wien, Austria
NHRS Naturhistorika Riksmusset, Sektionen för Entomologi, Stockholm, Sweden
NMEG Naturkundemuseum, Erfurt, Germany
NMPC National Museum, Prague, Czech Republic
NTUC National Taiwan University, Taipei, Taiwan, China
NYBM Brooklyn Museum, New York, USA
NZSI National Zoological Collection, Zoological Survey of India, Calcutta, India
OLML Oberösterreichisches Landesmuzeum, Linz, Austria
OSUO Oregon State University, Corvallis, Oregon, USA
OXUM Hope Entomological Collection, University Museum, Oxford, United Kingdom
PANS Academy of Natural Science, Philadelphia, USA
PMNH Connecicut, Peabody Museum of Natural History, New Haven, USA
QUST Qingdao University of Science and Technology, Qingdao, China
RIFID Research Institute of Forest Insect Diversity, Republic of Korea
RMNH Rijskmuseum van Natuurlijke Historie, Leiden, The Netherlands
SAMA South Australian Museum, Adelaide, Australia
SAMC Insect Collection, South African Museum, Cape Town, Republic of South Africa
SMTD Staatliches Museum für Tierkunde, Dresden, Germany
TCIS Technical Center of Animal-Plant & Food Interaction and Quarantine, Shenzen, China

UPPC	University of the Philippines, Laguna, Philippines
USNM	United States National Museum, Entomological Collection, Smithsonian Institution, Washington, USA
ZMAS	Zoological Museum, Russian Academy of Sciences, St. Petersburg, Russia
ZMUB	Museum für Naturkunde der Humboldt-Universität zu Berlin, Berlin, Germany
ZMUC	Department of Entomology, Zoological Museum, University of Copenhagen, Copenhagen, Denmark
ZMUH	Zoologische Institutbund Zoologische Museum, Universität von Hamburg, Hamburg, Germany
ZMUM	Zoological Museum, Lomonosov University, Moscow, Russia
ZRC	National University of Singapore, Raffles Museum of Biodiversity Research, Zoological Reference Collection, Singapore, China
ZSMC	Zoologischen Staatssammlung des Bayerischen Staates, Munich, Germany

New Nomenclatorical Acts

Genus *Bostrichoplectron* Geis, 2015 **syn. nov.** for *Bostrichoplites* Lesne, 1898
Rhyzopertha hordeum Matsumura in Hagstrum et Subramanyam, 2009 **syn. nov.** for *Rhizopertha dominica* (Fabricius, 1792)

Three species are newly combined, transferred from the genus *Bostrichoplectron* Geis, 2015
Bostrychoplites normandi (Lesne, 1897) **comb. nov.**
Bostrychoplites yemenensis Lesne, 1935 **comb. nov.**
Bostrychoplites zickeli (Marseul, 1867) **comb. nov.**

New homonyms – without designation of replacement name (see chapter Material and methods)
Bostrichus asperatus Ratzeburg, 1837 to *Bostrichus asperatus* Gyllenahl, 1813
Bostrichus ater Paykull, 1800 to *Bostrichus ater* Fabricius, 1792
Bostrichus elongatus Herbst, 1787 to *Bostrichus elongatus* Fabricius, 1782
Bostrichus ferrugineus Boheman, 1858 to *Bostrichus ferrugineus* Fabricius, 1801
Bostrichus pinastri Bechstein, 1818 to *Bostrichus pinastri* Bechstein in Bechstein et Sharfenberg, 1804
Bostrichus serratus Fabricius, 1801 to to *Bostrichus serratus* Panzer, 1795
Bostrichus thoracicus Fabricius, 1801 to *Bostrichus thoracicus* Panzer, 1796
Bostrichus xylographus C. R. Sahlberg, 1836 to *Bostrichus xylographus* Say, 1826

Catalogue

Bostrichiformia Forbes, 1926
Bostrichoidea Latreille, 1802

Family Bostrichidae Latreille, 1802
Bostrichini Latreille, 1802: 202
Type genus: Bostrichus Geoffroy, 1762
Bostrychidae Wollaston, 1867: 109 [LC]

†Subfamily Alitrepaninae Peng, Jiang, Engel et Wang, 2022

Alitrepaninae Peng, Jiang, Engel et Wang, 2022: 2
Type genus: Alitrepanum Peng, Jiang, Engel et Wang, 2022

†Genus *Poinarinius* Legalov, 2018
Poinarinius Legalov, 2018: 210
Type species: *Poinarinius burmaensis* Legalov, 2018
= *Alitrepanum* Peng, Jiang, Engel et Wang, 2022: 2
Type species: *Alitrepanum aladelicatum* Peng, Jiang, Shi, Song, Long, Engel et Wang, 2022

†*Poinarinius aladelicatum* (Peng, Jiang, Engel et Wang, 2022)
Alitrepanum aladelicatum Peng, Jiang, Shi, Song, Long, Engel et Wang, 2022: 3 [QUST]
Distribution: I: BM (Cretaceous, Burmese amber)

†*Poinarinius antonkozlovi* Legalov et Háva, 2022
Poinarinius antonkozlovi Legalov et Háva, 2022: 5 [ISEA]
Distribution: I: BM (Cretaceous, Burmese amber)

†*Poinarinius aristovi* Legalov et Háva, 2022
Poinarinius aristovi Legalov et Háva, 2022: 3 [ISEA]
Distribution: I: BM (Cretaceous, Burmese amber)

†*Poinarinius borowskii* Legalov et Háva, 2022
Poinarinius borowskii Legalov et Háva, 2022: 8 [ISEA]
Distribution: I: BM (Cretaceous, Burmese amber)

 | DOI:10.1163/9789004707917_002

†*Poinarinius burmaensis* Legalov, 2018
Poinarinius burmaensis Legalov, 2018: 2010 [ISEA]
Distribution: I: BM (Cretaceous, Burmese amber)

†*Poinarinius coziki* Háva et Legalov, 2023
Poinarinius coziki Háva et Legalov, 2023b: 272 [JHAC]
Distribution: I: BM (Cretaceous, Burmese amber)

†*Poinarinius cretaceus* Legalov et Háva, 2022
Poinarinius cretaceus Legalov et Háva, 2022: 11 [ISEA]
Distribution: I: BM (Cretaceous, Burmese amber)

†*Poinarinius decimus* Háva & Legalov, 2023
Poinarinius decimus Háva & Legalov, 2023a: 286 [JHAC]
Distribution: I: BM (Cretaceous, Burmese amber)

†*Poinarinius lesnei* Legalov et Háva, 2022
Poinarinius lesnei Legalov et Háva, 2022: 9 [ISEA]
Distribution: I: BM (Cretaceous, Burmese amber)

†*Poinarinius perkovskyi* Legalov et Háva, 2022
Poinarinius perkovskyi Legalov et Háva, 2022: 5 [ISEA]
Distribution: I: BM (Cretaceous, Burmese amber)

†*Poinarinius zahradniki* Legalov et Háva, 2022
Poinarinius zahradniki Legalov et Háva, 2022: 8 [ISEA]
Distribution: I: BM (Cretaceous, Burmese amber)

Subfamily Bostrichinae Latreille, 1802

Bostrichini Latreille, 1802: 202
Type genus: *Bostrichus* Geoffroy, 1762

Tribe Apatini Billberg, 1820

Apatides Billberg, 1820a: 47
Type genus: *Apate* Fabricius, 1775
= Apatides Billberg, 1920b: 394 [**HN**]
Type genus: *Apate*, Fabricius, 1775
= Apatini Liu et Schönitzer, 2011 (sic!) [**HN**]
= Apatini Liu, Laevengood et Bernal, 2022: 21 [**HN**]
Type genus: *Apate* Fabricius, 1775

= Bostrychopsini Lesne, 1921c: 288
Type genus: *Bostrychopsis* Lesne, 1899a
= Dinapatini Lesne, 1910: 471
Type genus: *Dinapate* Horn
= Ligniperdidae Jacobi, 1906: 139
Type genus: *Ligniperda* Pallas, 1772

Genus *Apate* Fabricius, 1775

Apate Fabricius, 1775: 54
Type species: *Apate terebrans* Pallas, 1772
= *Bostrychopsis* Lesne, 1899a: 524 [HN]
Type species: *Bostrichus cephalotes* Olivier, 1790
= *Ligniperda* Pallas, 1772: 7
Type species: *Ligniperda terebrans* Pallas, 1772

Apate bicolor Fåhraeus, 1871

Apate bicolor Fåhraeus, 1871: 665 [NHRS]
Distribution: E: CG CK GO KY MW MZ SF TZ ZA ZB ZI

Apate bilabiata Lesne, 1909

Apate bilabiata Lesne, 1909b: 535 [ZMUB]
Distribution: E: KY

Apate congener (Gerstäcker, 1855)

Ligniperda congener Gerstäcker, 1855: 268 [ZMUB]
= *Apate frontalis* Fåhraeus, 1871: 664 [NHRS]
= *Apate hirtifrons* Latreille in Dejean, 1835: 308 [MNHN?] [NN]
= *Ligniperda cylindricus* Gerstäcker, 1855: 268 [ZMUB]
Distribution: E: CG CM EQ ER GT KY MB MI MT MU MZ RI SF SO TZ UG ZA ZB; **Nt_i:** GL; **P_i:** EG

Apate degener Murray, 1867

Apate degener Murray, 1867: 87 [BMNH]
Distribution: E: AA BN CF CG CK GH GS GO GX IV KY LI MZ NX SR STI TZ UG ZA ZI

Apate ecomata Lesne, 1929

Apate ecomata Lesne, 1929b: 67 [MCSN]
Distribution: E: SO

Apate geayi Lesne, 1907

Apate geayi Lesne, 1907b: 325 [MNHN]
Distribution: E: MB; **P_i:** FR (Corsica)

Apate indistincta Murray, 1867

Apate monacha indistincta Murray, 1867: 88 [MNHN]

= *Apate anceps* Fåhraeus, 1871: 663 [NHRS]

= *Apate franscisca* Vrydagh, 1958c: 19 [unknown] [**NN**]

Distribution: **E:** AA CK DJ EH ER KY MB MU MZ SF SO TZ ZA

Apate lignicolor Fairmaire, 1883

Apate lignicolor Fairmaire, 1883a: 95 [MNHN]

= *Apate femoralis* Fåhraeus, 1871: 664 [NHRS] [HN]

Distribution: **E:** BD BN CG EH EQ GH GO IV KY MB ML MZ NI NX SF SO SSU SU TZ UG ZA ZB

Apate monachus Fabricius, 1775

Apate monachus Fabricius, 1775: 54 [BMNH?]

= *Apate anachoreta* Dejean, 1821: 101 [MNHN?] [**NN HN**]

= *Apate anachoreta* Dejean, 1835: 308 [MNHN?] [**NN**]

= *Apate carmelita* Fabricius, 1801: 379 [ZMUC]

= *Apate francisca* Fabricius, 1801: 379 [ZMUC]

= *Apate monachus* Dejean, 1821: 101 [MNHN?] [**NN**]

= *Apate monachus* Duftschidt in Dejean, 1821: 101 [MNHN?] [**NN HN**]

= *Apate monachus* Duftschidt in Dejean, 1835: 308 [MNHN?] [**NN**]

= *Apate monachus* Sturm, 1826: 95 [unknown] [**NN**]

= *Apate monachus* Olivier in Sturm, 1843: 231 **E:** GX [unknown] [**NN HN**]

= *Apate leprieurii* Buquet in Dejean, 1835: 308 [MNHN?] [**NN**]

= *Apate reticulata* Dejean, 1835: 308 [MNHN?] [**NN**]

= *Apate rufiventris* Lucas, 1843a: 159 [MNHN]

= *Apate semicostata* J. Thomson, 1858: 83 [MNHN]

= *Bostrichus mendicus* Olivier, 1790a: 108 [MNHN]

= *Bostrychus senni* Stefani-Perez, 1911: 62 [unknown]

= *Synodendron gibbum* Fabricius, 1798: 156 [ZMUC]

Distribution: **E:** AA CD CG CK CM$_i$ CV DJ EH EQ ER GH GO GS GX IV KY MAY MB MZ NI NX RW SF SG SO SR STI SU TZ UG ZA ZB; **N$_i$:** US; **Nt$_i$:** BZ (BZE) CU HA JC MX PR VE „Lesser Antilles Is.“; **P$_i$:** AB AG CY EG FR GB GE GG HU IN IS IT LB LE MA MO NL PT SP SV SY SZ TU YE (YEC)

Apate raricoma Lesne, 1924

Apate raricoma Lesne, 1924: 256 [ISNB MNHN]

Distribution: **E:** IV

Apate reflexa Lesne, 1909

Apate reflexa Lesne, 1909b: 533 [MNHN ISNB DEIC]

Distribution: **E:** CG CK EQ GO GX NX RW SR SU ZA

Apate scoparia Lesne, 1909

Apate scoparia Lesne, 1909b: 518 [MNHN BMNH MCSN]

Distribution: E: CK EH KY SO TZ UG

Apate subcalva Lesne, 1923

Apate subcalva Lesne, 1923a: 60 [MNHN ZMUB]

Distribution: E: CG EQ GO GX NX UG ZA

Apate submedia Walker, 1858

Apate submedia Walker, 1858b: 286 [BMNH]

= *Apate submedia* Walker, 1859: 260 [**HN**], [BMNH]

Distribution: I: CX ID (IDS) TH

Apate terebrans (Pallas, 1772)

Ligniperda terebrans Pallas, 1772: 7 [ZMUB]

= *Apate barbifrons* Dupont in Dejean, 1835: 308 [MNHN?] [**NN**]

= *Apate dispar* Fåhraeus, 1871: 663 [NHRS]

= *Apate hamatus* Fabricius, 1787: 33 [ZMUC?]

= *Apate terebrans* Olivier in Dejean, 1821: 101 [MNHN?] [**NN**]

= *Apate terebrans* Olivier in Dejean, 1835: 308 [MNHN?] [**NN**]

Distribution: E: AA BN CK EH EQ ER GX IV MB ML MZ NB NX SF SG SO STI SU TZ UG ZA ZB; **N_i:** US; **Nt_i:** AT BZ (BZC) CU JC; **P_i:** AB GB GG MO SP YE (YEC)

Apate Fabricius, 1775, other nomina nuda

Apate approximata Dupont in Dejean, 1835: 308 **E:** MB [MNHN?] [**NN**]

Apate armata Dejean, 1835: 309 **E:** SF [MNHN?] [**NN**]

Apate barbata Melsheimer, 1806: 7 [unknown] [**NN**]

Apate bicornis Dejean, 1835: 309 **N:** „America borealis“ [MNHN?] [**NN HN**]

Apate bicornis Melsheimer, 1806: 7 **N:** US (UST) [MCZC] [**NN**]

Apate bicornis Sturm, 1843: 231 **N:** „America borealis“ [unknown] [**NN HN**]

Apate bispinosa Dejean, 1821: 101 **Nt:** BZ [MNHN?] [**NN**]

Apate brunnipennis Billberg in Sturm, 1826: 94 [unknown] [**NN**]

Apate coenobita Dejean, 1821: 101 **Nt:** BZ [MNHN?] [**NN**]

Apate coenobita Dejean, 1835: 308 **Nt:** BZ [MNHN?] [**NN HN**]

Apate comolli Sturm, 1843: 231 **P:** IT [unknown] [**NN**]

Apate costata Sturm, 1826: 94 **Nt:** CU [unknown] [**NN**]

Apate cribraria Dejean, 1835: 308 **E:** GX [MNHN?] [**NN**]

Apate dominicana Eschscholtz in Dejean, 1835: 309 **I:** PH [MNHN?] [**NN**]

Apate gysselenii Dejean, 1835: 309 **P:** AU [MNHN?] [**NN**]

Apate hamaticollis Dejean, 1835: 308 **E:** SG [MNHN?] [**NN**]

Apate laevicollis Sturm, 1843: 231 **I:** ID occidentalis [unknown] [**NN**]
Apate longicornis Mannerheim in Dejean, 1835: 308 **Nt:** DR [MNHN?] [**NN**]
Apate longula Billberg in Sturm, 1826: 94 **P:** SV [unknown] [**NN**]
Apate longula Billberg in Sturm, 1843: 231 **I:** ID occidentalis [unknown] [**NN HN**]
Apate melanura Sturm, 1843: 231 **Nt:** CU [unknown] [**NN**]
Apate minuta Sturm, 1826: 95 **E:** SR [unknown] [**NN**]
Apate minutus Melsheimer, 1806: 7 **N:** US (UST) [MCZC] [**NN**]
Apate mucronata Sturm, 1843: 231 **N:** „America borealis" [unknown] [**NN**]
Apate nana Dejean, 1835: 309 **E:** MU [MNHN?] [**NN**]
Apate niger Melsheimer, 1806: 7 [unknown] [**NN**]
Apate obscura Sturm, 1826: 95 **Nt:** BZ [unknown] [**NN**]
Apate obscura Sturm, 1843: 231 **Nt:** BZ [unknown] [**NN HN**]
Apate perforata Schönherr in Dejean, 1835: 308 **E:** GX [MNHN?] [**NN**]
Apate perplexa Dejean, 1835: 308 **Nt:** BZ [MNHN?] [**NN**]
Apate politus Melsheimer, 1806: 7 [unknown] [**NN**]
Apate punctata Dejean, 1835: 309 **N:** „America borealis" [MNHN?] [**NN**]
Apate quadrispinosa Dejean, 1835: 309 **E:** SG [MNHN?] [**NN**]
Apate retusa Sturm, 1826: 95 **I:** ID orientalis [unknown] [**NN**]
Apate retusa Sturm, 1843: 231 **I:** ID occidentalis [unknown] [**NN HN**]
Apate rufescens Dejean, 1835: 309 **Nt:** BZ [MNHN?] [**NN**]
Apate ruficornis Dejean, 1835: 309 **E:** SF [MNHN?] [**NN**]
Apate scabricollis Sturm, 1826: 95 **I:** ID orientalis [unknown] [**NN**]
Apate scabricollis Sturm, 1843: 231 **I:** ID occidentalis. [unknown] [**NN HN**]
Apate scabricollis Sturm, 1843: 231 [unknown] [**NN HN**]
Apate serricollis Sturm, 1843: 231 [unknown] [**NN**]
Apate sordida Dejean, 1835: 309 **N:** „America borealis" [MNHN?] [**NN**]
Apate striatus Melsheimer, 1806: 7 [unknown] [**NN**]
Apate subdentata Dejean, 1835: 309 **N:** „America borealis" [MNHN?] [**NN**]
Apate thoracicornis Dejean, 1835: 309 **P:** EG [MNHN?] [**NN**]
Apate thoracica Sturm, 1826: 95 **I:** India occidentalis [unknown] [**NN**]
Apate thoracica Sturm, 1843: 231 **Nt:** BZ [unknown] [**NN HN**]
Apate thoracica Sturm, 1843: 231 **I:** ID occidentalis [unknown] [**NN HN**]
Apate truncata Latreille in Dejean, 1835: 309 **E:** MU [MNHN?] [**NN**]
Apate tuberculosa Gistel, 1848: 132 **Nt:** BZ [MNHN?] [**NN**]
Apate vittatus Melsheimer, 1806: 7 **N:** US (UST) [MCZC] [**NN**]
Apate westermanni Dejean, 1835: 308 **E:** GX [MNHN?] [**NN**]

Genus *Dinapate* Horn, 1886

Dinapate Horn, 1886: 1
Type species: *Dinapate wrightii* Horn, 1886

Dinapate hughleechi Cooper, 1986

Dinapate hughleechi Cooper, 1986: 83 [CASC FSCA]

Distribution: Nt: MX (MXE MXW)

Dinapate wrightii Horn, 1886

Dinapate wrightii Horn, 1886: 2 [MCZC]

Distribution: N: MX (MXC) US (USC)

Genus *Phonapate* Lesne, 1895

Phonapate Lesne, 1895a: 178

Type species: *Apate uncinata* Karsch, 1881

= *Megabostrichus* Chûjô, 1964: 206

Type species: *Megabostrichus imadatei* Chûjô, 1964

Phonapate andriana Lesne, 1909

Phonapate andriana Lesne, 1909b: 565 [MNHN ZMUB]

Distribution: E: MB

Phonapate chan (Semenow, 1891)

Apate chan Semenow, 1891: 351 [ZMAS]

Distribution: P: KZ TM UZ

Phonapate densoana Lesne, 1934

Phonapate madecassa densoana Lesne, 1934c: 42 [MNHN SMTD]

Distribution: E: MB

Phonapate deserti (Semenow, 1891)

Apate deserti Semenow, 1891: 351 [ZMAS]

Distribution: P: KZ TM UZ

Phonapate discreta Lesne, 1906

Phonapate discreta Lesne, 1906f: 415 [MCSN]

Distribution: E: EQ STI

Phonapate fimbriata Lesne, 1909

Phonapate fimbriata Lesne, 1909b: 568 [MNHN ZMUB NMNH]

Distribution: I: ID IA (IAC) LO MY (MYC) TH VT P: CH (SE SW)

Phonapate madecassa Lesne, 1899

Phonapate madecassa Lesne, 1899c: 10 [MNHN]

Distribution: E: MB

Phonapate nitidipennis (Waterhouse, 1881)

Phonapate nitidipennis Waterhouse, 1881: 472 [BMNH]

= *Apate uncinata* Karsch, 1881: 46 [ZMUB]

= *Ligniperda ganglbaueri* Zoufal, 1894: 37 [NHMW]

= *Phonapate frontalis arabs* Lesne, 1909b: 555 [MNHN MCSN BMNH NZSI MCZR]

= *Phonapate frontalis moghrebica* Lesne, 1934b: 218 [MNHN]

= *Phonapate uncinata* Lesne, 1902a: 123 [HN]

= *Phonapate uncinata africana* Vrydagh, 1961b: 9 [MNHN]

Distribution: E: CD DJ EH EQ ER KY MT MZ SF SG SO SU TZ ZB; **I_i:** ID; **Nt_i:** GL; **P:** AE AF AG BA CY EG GE_i IN IQ IS LB MO OM OP PA QA SA SP TU TR YE (YEC YES)

Phonapate porrecta Lesne, 1900

Phonapate porrecta Lesne, 1900b: 426 [ISNB MNHN]

Distribution: E: AA BN CG GO IV NB NX TG TZ TZ (Zanzibar) UG ZA

Phonapate stridula Lesne, 1909

Phonapate stridula Lesne, 1909b: 549 [MNHN BMNH]

= *Megabostrichus imadatei* Chûjô, 1964: 208 [KUFA]

Distribution: I: BM IA (IAC) ID TH VT; **P:** CH (SW)

Phonapate sublobata Lesne, 1909

Phonapate sublobata Lesne, 1909b: 570 [ZMUB]

Distribution: I: IA (IAS)

†Genus *Protapate* Wickham, 1912

Protapate Wickham, 1912: 20

Type species: *Protapate contorta* Wickham, 1912

†***Protapate contorta*** Wickham, 1912

Protapate contorta Wickham, 1912: 20 [PMNH]

Distribution: N: US (USU – Miocene, Florissant)

Genus *Xylomedes* Lesne, 1902

Xylomedes Lesne, 1902a: 118

Type species: *Apate rufocoronata* Fairmaire, 1892a

Xylomedes bidasi Borowski et Lasoń in Borowski, Lasoń et Sławski, 2021

Xylomedes bidasi Borowski et Lasoń in Borowski, Lasoń et Sławski, 2021: 292 [DFPC]

Distribution: P: IN

Xylomedes carbonnieri (Lesne, 1897)

Apate carbonnieri Lesne, 1897a: 235 [MNHN]

Distribution: **P:** AG LB TU

Xylomedes cornifrons cornifrons (Baudi di Selve, 1874)

Apate cornifrons Baudi di Selve, 1874: 334 [BMNH]

= *Xylomedes turcica* Lesne, 1941a: 134 [BMNH]

Distributon: **P:** CY IQ IS LE SY TR YE (YEC)

Xylomedes cornifrons rufocoronata (Fairmaire, 1892)

Apate rufocoronata Fairmaire, 1892a: 104 [MNHN]

Distribution: **E:** DJ EH ER KY SO SU; **P:** AG EG LB MO OM SA TU YE (YEC)

Xylomedes coronata (Marseul, 1883)

Apate coronata Marseul, 1883: 183 [MNHN]

= *Apate sericans* Marseul, 1883: 183 [MNHN]

Distribution: **P:** AG EG IS LB MO TU

Xylomedes laticornis (Lesne, 1895)

Apate laticornis Lesne, 1895a: 178 [MNHN]

Distribution: **E:** DJ EH ER; **P:** EG IN SA YE (YEC)

Xylomedes scutifrons Lesne, 1908

Xylomedes scutifrons Lesne, 1908d: 427 [MNHN]

Distribution: **E:** NB TZ

Tribe Bostrichini Latreille, 1802

Bostrichini Latreille, 1802: 202

Type genus: *Bostrichus* Geoffroy, 1762

= Apatidini Bradley, 1930: 207

Type genus: *Apatides* Casey, 1898

= Bostrichini Liu & Schönitzer, 2011 (sic!) = Bostrichini Liu, Leavengood et Bernal, 2022: 23 [**HN**]

= Lichenophanini Portevin, 1931: 470

Type genus: *Lichenophanes* Lesne, 1899

Genus *Amphicerus* LeConte, 1861

Amphicerus LeConte, 1861: 208

Type species: *Apate bicaudatus* Say, 1824

= *Caenophrada* Waterhouse, 1888: 350

Type species: *Caenophrada anobioides* Waterhouse, 1888

= *Schistoceros* Lesne, 1899a: 502

Type species: *Bostrichus bimaculatus* Olivier, 1790

Amphicerus anobioides (Waterhouse, 1888)

Caenophrada anobioides Waterhouse, 1888: 350 [BMNH]

Distribution: **E:** DJ ER; **I:** BM CX DJ ID (IDC IDW) LO TH; **P:** AF ID (UP) NP PA SA YE (YEC)

Amphicerus bicaudatus (Say, 1824)

Apate bicaudatus Say, 1824: 320 [USNM]

= *Apate aspericollis* Germar, 1824: 465 [ZMUB] [**NO**]

= *Apate aspericollis* Sturm, 1843: 231 [unknown] [**NN**]

= *Apate cornutus* Melsheimer, 1806: 7 N: US (UST) [MCZC] [**NN**]

= *Apate cornutus* Melsheimer in Say, 1824: 320 [**NN**]

= *Apate cornutus* Melsheimer in Sturm, 1843: 231 [**NN**]

= *Amphicerus gracilis* Casey, 1898: 69 [USNM]

= *Apate modesta* Dejean, 1835: 309 [MNHN?] [**NN**]

= *Apate serricollis* Germar, 1824: 464 [ZMUB] [**NO**]

Distribution: **N:** CN US (USD USE); **Nt_i:** CU GL MX; **P_i:** GE

Amphicerus bimaculatus (Olivier, 1790)

Bostrichus bimaculatus Olivier, 1790a: 109 [MNHN?]

= *Apate aurita* Frivaldszky von Frivald, 1835: 267 [HNMH?]

= *Apate bimaculatus* Dejean, 1821: 101 [MNHN?] [**NN**]

= *Apate bimaculatus* Fabricius, 1792: 362 [ZMUC?] [**HN**]

= *Apate bimaculatus* Sturm, 1826: 94 [MNHN?] [**NN HN**]

Distribution: **N_i:** US; **Nt_i:** UR; **P:** AB AF AG AL BH BU CR CY FR GE GR HU IN IS IT KY KZ MA MC MO PA PT RO RU (ST) SP SY TD TM TR TU UK YU

Amphicerus caenophradoides (Lesne, 1895)

Bostrychus caenophradoides Lesne, 1895a: 174 [ISNB MNHN]

= *Bostrychus malayanus* Lesne, 1898a: 255 [RNMH MNHN]

Distribution: **I:** CA IA (IAB IAJ IAS) LO MY (MYC) TH VT

Amphicerus clunalis Lesne, 1939

Amphicerus clunalis Lesne, 1939b: 100 [MNHN]

Distribution: **Nt:** MX (MXV)

Amphicerus cornutus (Pallas, 1772)

Ligniperda cornuta Pallas, 1772: 8 [ZMUB?]

= *Amphicerus maritimus* Casey, 1898: 68 [USNM]

= *Apate deflexa* Gorham, 1883: 213 [NN]

= *Apate deflexa* Sturm, 1843: 231 [NN]

= *Apate mexicana* Höpfner in Dejean, 1835: 308 [MNHN?] [NN]

= *Apate punctipennis* LeConte, 1858: 73 [MCZC]

= *Apate tristis* Gorham, 1883: 213 [NN]

= *Apate tristis* Sturm, 1843: 231 [NN]

= *Bostrichus bicornutus* Latreille, 1813: 65 [MNHN?]

= *Bostrichus migrator* Sharp in Blackburn & Sharp, 1885: 160 [BMNH]

= *Bostrichus peregrinus* Erichson, 1847: 87 [ZMUB]

= *Schistoceros consanguineus* Lesne, 1899a: 513 [MNHN]

Distribution: **A_i:** FJ HI; **E_i:** STI; **N:** CN (CNN) MX (MXC) US (USE USO); **Nt:** AT AY (AYN) BI BR BV BZ (BZC BZS) CB CC CL (CLN) CMS CU DO EC (EC GI) GL GN GT HO JC LW MNT MQ MX (MXT MXW) NG NH PA PE PG PN PR SN SQ STB STT VE VI; **P_i:** PL

Amphicerus galapaganus Lesne, 1910

Schistoceros cornutus galapaganus Lesne, 1910c: 184 [BMNH]

Distribution: **Nt:** GI

Amphicerus lignator (Lesne, 1899)

Schistoceros lignator Lesne, 1899a: 509 [NHMW]

Distribution: **Nt:** CB VE

Amphicerus securimentum Lesne, 1939

Amphicerus securimentum Lesne, 1939b: 99 [DEIC]

Distribution: **Nt:** MX (MXT)

Amphicerus simplex (Horn, 1885)

Sinoxylon simplex Horn, 1885: 155 [PANS MNHN]

= *Amphicerus brevicollis* Casey, 1898: 70 [USNM]

= *Amphicerus grandicollis* Casey, 1898: 69 [USNM]

Distribution: **N:** US (USO)

†*Amphicerus sublaevis* Wickham, 1914

Amphicerus sublaevis Wickham, 1914: 452 [MCZC]

Distribution: **N:** US (USU – Miocene, Florissant)

Amphicerus teres Horn, 1878

Amphicerus teres Horn, 1878: 547 [PANS]

Distribution: N: US (USC USO)

Amphicerus tubularis (Gorham, 1883)

Xylopertha tubularis Gorham, 1883: 216 [BMNH]

Distribution: Nt: MX (MXM MXT) PN

Genus *Apatides* Casey, 1898

Apatides Casey, 1898: 66

Type species: *Amphicerus fortis* LeConte, 1866

Apatides fortis (LeConte, 1866)

Amphicerus fortis LeConte, 1866: 101 [MCZC]

= *Apatides pallens* Casey, 1914: 359 [USNM]

= *Apatides puncticeps* Casey, 1898: 71 [USNM]

= *Apatides robustus* Casey, 1898: 71 [USNM]

Distribution: N: MX (MXC) US (USC USO USU); **Nt:** CB CC NG

Genus *Bostrichus* Geoffroy, 1762

Bostrichus Geoffroy, 1762: 301

Type species: *Dermestes capucinus* Linnaeus, 1758

= *Bostrychus* Agassiz, 1846: 49 [LC]

Bostrichus capucinus (Linnaeus, 1758)

Dermestes capucinus Linnaeus, 1758: 355 [unknown]

= *Apate nigriventris* Lucas, 1843a: 159 [MNHN]

= *Apate rugosa* Fabricius, 1801: 380 [ZMUC]

= *Bostrichus capucinus* Dejean, 1821: 101 [MNHN?] [NN]

= *Bostrichus capucinus parvulus* Gemminger & Harold, 1869: 1791 [NN]

= *Bostrychus capucinus rubriventris* Zoufal, 1894: 40 [unknown]

= *Bostrichus luctuosus* Olivier, 1790a: 109 [MNHN]

= *Bostrichus ruber* Geoffroy in Fourcroy, 1785: 133 [unknown]

Distribution: E_i: SU TG; N_i: US (UST); **P:** AB AG AL AR AU BE BH BU BY CR CY CZ DE EG FI FR GB GE GG GR HU CH (CE NO NW SW) ID (KA) IN IQ IR IS IT JO KY KZ LA LB LE LT MA MC MD ME MO NL NR PL PT RO RU (CT ES NT ST WS) SB SK SL SP SV SY SZ TM TR TU UK UZ

Bostrichus Geoffroy, 1762, other nomina nuda

Bostrichus badius Dupont in Dejean, 1835: 307 **E:** MB [MNHN?] [NN]

Bostrichus bidens Dejean, 1821: 101 **P:** SV [MNHN?] [NN]

Bostrichus bidens Sturm, 1826: 101 **P:** GE [unknown] [**NN HN**]
Bostrichus bispinus Fabricius in Sturm, 1843: 230 **P:** GE [unknown] [**NN**]
Bostrichus bispinus Megerle in Dejean, 1821: 101 **P:** AU [MNHN?] [**NN**]
Bostrichus bispinus Megerle in Dejean, 1835: 307 **P:** AU [MNHN?] [**NN HN**]
Bostrichus bispinus Megerle in Sturm, 1826: 101 **P:** AU [unknown] [**NN HN**]
Bostrichus bispinus Megerle in Sturm, 1843: 230 **P:** AU [unknown] [**NN HN**]
Bostrichus brunneus Sturm, 1843: 230 **N:** „America borealis“ [unknown] [**NN**]
Bostrichus brunnipes Sturm, 1843: 230 **N:** „America borealis“ [unknown] [**NN**]
Bostrichus calcaratus Dejean, 1821: 101 **P:** CR [MNHN?] [**NN**]
Bostrichus calcaratus Dejean, 1835: 307 **P:** CR [MNHN?] [**NN HN**]
Bostrichus castaneus Sturm, 1826: 230 **P:** AU [unknown] [**NN**]
Bostrichus castaneus Sturm, 1843: 102 **P:** GE [unknown] [**NN HN**]
Bostrichus cavicollis Lacordaire in Dejean, 1835: 307 **Nt:** FG [MNHN?] [**NN**]
Bostrichus chalcographus Dejean, 1821: 101 **P:** SV [MNHN?] [**NN HN**]
Bostrichus chalcographus Fabricius in Dejean, 1835: 307 **P:** SV [MNHN?] [**NN HN**]
Bostrichus chalcographus Sturm, 1826: 101 **P:** SV [unknown] [**NN HN**]
Bostrichus chalcographus Ziegler in Dejean, 1835: 307 **P:** AU [MNHN?] [**NN HN**]
Bostrichus chloroticus Schönherr in Dejean, 1835: 307 **Nt:** JC [MNHN?] [**NN**]
Bostrichus chloroticus Schönherr in Sturm, 1843: 230 **N:** „America borealis“ [MNHN?] [**NN HN**]
Bostrichus cinereus Gyllenhal in Dejean, 1821: 101 **P:** SV [MNHN?] [**NN**]
Bostrichus cinereus Gyllenhal in Dejean, 1835: 308 **P:** SV [MNHN?] [**NN HN**]
Bostrichus cinereus Gyllenhal in Sturm, 1826: 101 **P:** SV [unknown] [**NN HN**]
Bostrichus comosus Sturm, 1826: 101 **P:** GE [unknown] [**NN**]
Bostrichus comosus Sturm, 1843: 230 **P:** GE [unknown] [**NN HN**]
Bostrichus conformis Dejean, 1835: 307 **N:** „America borealis“ [MNHN?] [**NN**]
Bostrichus crinitus Sturm, 1843: 230 **Nt:** „America borealis“ [unknown] [**NN**]
Bostrichus denticulatus Sturm, 1826: 101 **P:** GE [unknown] [**NN**]
Bostrichus denticulatus Sturm, 1843: 230 **N:** „America borealis“ [unknown] [**NN HN**]
Bostrichus discicollis Sturm, 1843: 230 **N:** „America borealis“ [unknown] [**NN**]
Bostrichus dispar Dejean, 1821: 101 **P:** FR [MNHN?] [**NN**]
Bostrichus elongatus Gysselen in Dejean, 1835: 307 **P:** AU [MNHN?] [**NN**]
Bostrichus exiguus Dejean, 1835: 308 **Nt:** BZ [MNHN?] [**NN**]
Bostrichus fagi Ziegler in Dejean, 1821: 101 **P:** CR [MNHN?] [**NN**]
Bostrichus fagi Ziegler in Dejean, 1835: 308 **P:** CR [MNHN?] [**NN HN**]
Bostrichus fagi Ziegler in Sturm, 1826: 101 **P:** GE [unknown] [**NN HN**]
Bostrichus fagi Ziegler in Sturm, 1843: 230 **P:** CR [unknown] [**NN HN**]
Bostrichus femoratus Lacordaire in Dejean, 1835: 307 **Nt:** FG [MNHN?] [**NN**]
Bostrichus ferrugineus Fabricius in Sturm, 1843: 230 **Nt:** CU [unknown] [**NN HN**]
Bostrichus ferrugineus Sturm, 1826: 101 **P:** GE [unknown] [**NN**]
Bostrichus fici Dejean, 1821: 101 **P:** CR [MNHN?] [**NN**]

Bostrichus fici Dejean, 1835: 308 **P:** FR [MNHN?] [**NN HN**]
Bostrichus fuscus Gyllenhal in Dejean, 1835: 307 **P:** „P“ [MNHN?] [**NN**]
Bostrichus fuscus Marsham in Sturm, 1843: 230 **P:** AU [unknown] [**NN HN**]
Bostrichus gilvipes Sturm, 1843: 230 **N:** „America borealis“ [unknown] [**NN**]
Bostrichus iconographus Kugelann in Dejean, 1835: 307 **P:** RU [MNHN?] [**NN**]
Bostrichus impressus Gravenhorst in Sturm, 1826: 101 **P:** GE [unknown] [**NN**]
Bostrichus impressus Gravenhorst in Sturm, 1843: 230 **P:** AU [unknown] [**NN HN**]
Bostrichus laevigatus Dejean, 1835: 307 **N:** „America borealis“ [MNHN?] [**NN**]
Bostrichus laricis denticulatus Sturm in Dejen, 1833: 307 **P:** GE [MNHN?] [**NN HN**]
Bostrichus minutissimus Dejean, 1835: 308 **N:** „America borealis“ [MNHN?] [**NN**]
Bostrichus minutus Dejean, 1835: 307 **Nt:** BZ [MNHN?] [**NN HN**]
Bostrichus minutus Duftschmid in Dejean, 1821: 101 **P:** SV [MNHN?] [**NN**]
Bostrichus minutus Duftschmid in Dejean, 1835: 308 **P:** AU [MNHN?] [**NN HN**]
Bostrichus monographus Dejean, 1821: 101 **P:** SV [MNHN?] [**NN**]
Bostrichus monographus Schönherr in Dejean, 1821: 101 **P:** SV [MNHN?] [**NN HN**]
Bostrichus monographus Schönherr in Dejean, 1835: 307 **P:** SV [MNHN?] [**NN HN**]
Bostrichus monographus Sturm, 1826: 101 **P:** GE [unknown] [**NN HN**]
Bostrichus multidentatus Sturm, 1826: 101 **N:** „America borealis“ [unknown] [**NN**]
Bostrichus multidentatus Sturm, 1843: 230 **N:** „America borealis“ [unknown] [**NN HN**]
Bostrichus nanus Dejean, 1835: 308 **P:** EG [MNHN?] [**NN**]
Bostrichus nigritus Dejean, 1835: 307 **P:** SV [MNHN?] [**NN**]
Bostrichus obscurus Dejean, 1821: 101 **P:** AU [MNHN?] [**NN**]
Bostrichus obscurus Dejean, 1835: 307 **P:** AU [MNHN?] [**NN HN**]
Bostrichus parvulus Dejean, 1835: 307 **N:** „America borealis“ [MNHN?] [**NN**]
Bostrichus perforatus Gyllenhal in Sturm, 1826: 101 **P:** SV [unknown] [**NN**]
Bostrichus pilosulus Linz in Sturm, 1826: 101 **P:** GE [unknown] [**NN**]
Bostrichus pilosulus Sturm, 1843: 230 **P:** GE [unknown] [**NN HN**]
Bostrichus polygraphus Dejean, 1821: 101 **P:** SV [MNHN?] [**NN**]
Bostrichus polygraphus Fabricius in Dejean, 1835: 307 **P:** SV [MNHN?] [**NN HN**]
Bostrichus polygraphus Sturm, 1826: 101 **P:** SV [unknown] [**NN HN**]
Bostrichus porcatus Dejean, 1821: 101 **P:** CR [MNHN?] [**NN**]
Bostrichus porcatus Dejean, 1835: 308 **P:** CR [MNHN?] [**NN HN**]
Bostrichus porographus Eschscholtz in Dejean, 1835: 307 **P:** RU [MNHN?] [**NN**]
Bostrichus porographus Eschscholtz in Sturm, 1843: 230 **P:** „Lapponia] [unknown] [**NN HN**]
Bostrichus praeustus Dejean, 1835: 307 **N:** „America borealis“ [MNHN?] [**NN**]
Bostrichus pubescens Lacordaire in Dejean, 1835: 307 **Nt:** FG [MNHN?] [**NN**]
Bostrichus pubescens Steven in Dejean, 1835: 307 **P:** RU [MNHN?] [**NN HN**]
Bostrichus pubescens Steven in Sturm, 1843: 230 **P:** GE [unknown] [**NN HN**]
Bostrichus pubescens Sturm, 1843: 230 **P:** AU [unknown] [**NN HN**]
Bostrichus pulicarius Dejean, 1835: 308 **N:** „America borealis“ [MNHN?] [**NN**]

Bostrichus pumilus Dejean, 1835: 308 **N:** „America borealis“ [MNHN?] [**NN**]
Bostrichus punctatus Chevrolat in Dejean, 1835: 307 **Nt:** CB [MNHN?] [**NN**]
Bostrichus pygmaeus Dejean, 1835: 307 **E:** MU [MNHN?] [**NN**]
Bostrichus quadridentatus Sturm, 1826: 101 **P:** GE [unknown] [**NN HN**]
Bostrichus quadridentatus Sturm, 1843: 230 **P:** „Lapponia“ [unknown] [**NN**]
Bostrichus rufipes Latreille in Dejean, 1835: 307 **P:** FR [MNHN?] [**NN**]
Bostrichus seminarius Schönherr in Dejean, 1835: 307 **I:** PH [MNHN?] [**NN**]
Bostrichus septentrionis LeConte in Dejean, 1835: 307 **N:** „America borealis“ [MNHN?] [**NN**]
Bostrichus serraticollis Ullrich in Sturm, 1843: 230 **P:** AU [unknown] [**NN**]
Bostrichus suturalis Dejean, 1821: 101 **P:** SV [MNHN?] [**NN**]
Bostrichus suturalis Dejean, 1835: 307 **P:** SV [MNHN?] [**NN HN**]
Bostrichus tiliae Fabricius in Dejean, 1835: 308 **P:** SV [MNHN?] [**NN**]
Bostrichus tiliae Gyllenhal in Sturm, 1826: 101 **P:** GE [unknown] [**NN HN**]
Bostrichus troglodytes Dejean, 1835: 308 **P:** CR [MNHN?] [**NN**]
Bostrichus ustulatus Sturm, 1843: 230 **N:** „America borealis“ [unknown] [**NN**]
Bostrichus vicinus Dejean, 1835: 307 **N:** „America borealis“ [MNHN?] [**NN**]
Bostrichus villosus Sturm in Dejean, 1821: 101 **P:** AU [MNHN?] [**NN**]
Bostrichus villosus Sturm in Dejean, 1821: 101 **P:** „P“ [MNHN?] [**NN HN**]
Bostrichus villosus Gyllenhal in Dejean, 1835: 307 **P:** SV [MNHN?] [**NN HN**]
Bostrichus villosus Gyllenhal in Sturm, 1843: 230 **P:** GE [unknown] [**NN HN**]
Bostrichus xanthopus Dejean, 1835: 307 **E:** SF [MNHN?] [**NN**]

Genus *Bostrycharis* Lesne, 1925

Bostrycharis Lesne, 1925a: 25
Type species: *Bostrycharis niveosquamosus* Lesne, 1925a

Bostrycharis niveosquamosus Lesne, 1925

Bostrycharis niveosquamosus Lesne, 1925a: 26 [MNHN BMNH]
Distribution: **E:** LH MZ TZ ZB ZI

Genus *Bostrychoplites* Lesne, 1899

Bostrychoplites Lesne, 1899a: 565
Type species: *Bostrichus cornutus* Olivier, 1790
= *Bostrichoplectron* Geis, 2015: 75 **syn. nov.**
Type species: *Bostrychus normandi* Lesne, 1897

Bostrychoplites arabicus Lesne, 1935

Bostrychoplites arabicus Lesne, 1935f: 271 [ZMUH MNHN ZMUM]
Distribution: **P:** YE (YEC)

Bostrychoplites armatus Lesne, 1899

Bostrychoplites armatus Lesne, 1899a: 579 [MNHN]

Distribution: E: MB MZ MU SF ZA

Bostrychoplites cornutus (Olivier, 1790)

Bostrichus cornutus Olivier, 1790a: 108 [MNHN]

= *Apate armata* Quedenfeldt, 1887: 325 [NN]

= *Apate destructor* Burchell in Dejean, 1821: 101 [MNHN?] [**NN**]

= *Apate destructor* Burchell in Dejean, 1835: 309 [MNHN?] [**NN HN**]

= *Apate erosipennis* Buquet in Dejean, 1835: 309 [MNHN?] [**NN**]

= *Bostrychoplites megaceros* Lesne, 1899a: 574 [MNHN]

= *Bostrichus abyssinicus* Murray, 1867: 90 [BMNH]

= *Bostrichus cephalotes* Olivier, 1790a: 108 [MNHN]

= *Apate cornutus* Dejean, 1821: 101 [MNHN?] [**NN**]

= *Apate cornutus* Fabricius in Dejean, 1835: 309 [MNHN?] [**NN HN**]

= *Bostrychus cornutus minor* Quedenfeldt, 1887: 325 [NN]

Distribution: **A$_i$:** AS; **E:** AA BD BW CD CG CK CM EH EQ ER GA GO GS GX IV KY MB ML MU MW MZ NX RI RW SF SG SO SR STI SU SWA TZ UG ZA ZB ZI; **I:** ID; **N$_i$:** CN (CNM) US (USG); **Nt$_i$:** CB GL MX; **P:** AU$_i$ EG$_i$ FI$_i$ GB$_i$ GE$_i$ IT$_i$ MO$_i$ SA SP$_i$ SV$_i$ SZ$_i$ TR YE (YEC)

Bostrychoplites cylindricus (Fåhraeus, 1871)

Bostrichus cylindricus Fåhraeus, 1871: 668 [NHRS]

Distribution: E: AA BD CF CG CK EH KY MW MZ RW SF TG TZ UG ZA ZB ZI

Bostrychoplites dicerus Lesne, 1899

Bostrychoplites dicerus Lesne, 1899a: 580 [MNHN]

= *Bostrycholites tastei* Lesne, 1899a: 580 [**NN**]

Distribution: E: BF BN CG GX IV ML NI NX SG SSU TZ ZA

Bostrychoplites guineanus Lesne, 1923

Bostrychoplites guineanus Lesne, 1923a: 58 [MNHN MRAC ISNB]

= *Bostrychoplites guineanus orientalis* Basilewsky, 1955: 140 [MRAC]

Distribution: E: CG GA GH GS GX IV LI NX RW SR TZ ZA

Bostrychoplites luniger (J. Thomson, 1858)

Apate lunigera J. Thomson, 1858: 83 [MNHN]

= *Bostrichus brevicornutus* Murray, 1867: 91 [BMNH]

= *Bostrichus protrudens* Murray, 1867: 88 [BMNH]

Distribution: E: AA BN CG EQ GO IV NX TZ ZA

Bostrychoplites normandi (Lesne, 1897) **comb. nov.**

Bostrychus normandi Lesne, 1897a: 236 [MNHN]

Distribution: P: AG EG OM TR TU

Bostrychoplites peltatus Lesne, 1899

Bostrychoplites peltatus Lesne, 1899a: 580 [MNHN]

= *Bostrychoplites suturalis* Lesne, 1931b: 24 [MNHN]

Distribution: E: BW MB$_i$ NB SF TZ ZB ZI

Bostrychoplites productus (Imhoff, 1843)

Apate producta Imhoff, 1843: 176 [NHMB]

= *Apate ludovici* Fairmaire, 1883b: 133 [MNHN]

Distribution: E: AA BN CG CK EQ GG GH GS GX GO IV LI NX TG TZ UG ZA

Bostrychoplites valens Lesne, 1899

Bostrychoplites valens Lesne, 1899a: 578 [MNHN]

Distribution: E: AA BW NB SF TZ ZA ZI

Bostrychoplites vernicatus Lesne, 1923

Bostrychoplites vernicatus Lesne, 1923a: 58 [MNHN DEIC]

Distribution: E: CG CK GO GX IV NX ZA

Bostrychoplites yemenensis Lesne, 1935 **comb. nov.**

Bostrichoplites normandi yemenensis Lesne, 1935f: 272 [ZMUM]

Distribution: P: OM YE (YEC)

Bostrychoplites zickeli (Marseul, 1867) **comb. nov.**

Apate zickeli Marseul, 1867: 34 [MNHN]

= *Apate hamaticollis* Fairmaire, 1874: 407 [MNHN]

Distribution: E: CD DJ EH ER ML MT NX$_i$ SG SO SU; **P:** AG EG GE$_i$ LB MO OM SA TU YE (YEC)

Genus *Bostrychopsis* Lesne, 1899

Bostrychopsis Lesne, 1899a: 444

Type species: *Bostrychopsis villosula* Lesne, 1905

= *Dominikia* Borowski, 2020: 20 [NN]

= *Dominikia* Borowski & Węgrzynowicz, 2007a: 87

Type species: *Bostrychus parallelus* Lesne, 1895

Bostrychopsis bengalensis (Lesne, 1895)

Bostrychus bengalensis Lesne, 1895a: 174 [MNHN ISNB]

Distribution: I: ID (IDB IDP); **P:** ID (UP) PA

Bostrychopsis bozasi Lesne, 1913
Bostrychopsis bozasi Lesne, 1913e: 473 [MNHN MCSN]
Distribution: E: EH KY MB SO ZB

Bostrychopsis confossa (Fairmaire, 1880)
Apate confossa Fairmaire, 1880b: 308 [MNHN]
Distribution: E: MB

Bostrychopsis crinita Lesne, 1935
Bostrychopsis crinita Lesne, 1935a: 9 [ISNB]
= *Bostrychopsis crinita* Lesne in Collart, 1934: 244 [**NN**]
Distribution: E: CG SSU SU TZ ZA

Bostrychopsis cristaticeps Lesne, 1906
Bostrychopsis cristaticeps Lesne, 1906a: 403 [BMNH]
Distribution: P: YE (YES)

Bostrychopsis delkeskampi Vrydagh, 1959
Bostrychopsis delkeskampi Vrydagh, 1959c: 5 [ZMUB ISNB]
Distribution: E: KY

Bostrychopsis freyi Vrydagh, 1959
Bostrychopsis freyi Vrydagh, 1959c: 6 [NHMB HNHM]
Disribution: Nt: BZ (BZA BZS) PG

Bostrychopsis ganglbaueri Lesne, 1899
Bostrychopsis ganglbaueri Lesne, 1899a: 544 [MNHN NHMW]
= *Apate mutica* Dejean, 1835: 308 [MNHN?] [**NN**]
Distribution: Nt: BZ (BZC)

Bostrychopsis jesuita (Fabricius, 1775)
Apate jesuita Fabricius, 1775: 54 [BMNH]
= *Apate canarii* Nördlinger, 1880: 66 [unknown]
Distribution: A: AS (ASE ASN ASQ AST ASW) IA (IAN) PW; E_i: ZA; Nt_i: BZ; P_i: GE

Bostrychopsis laminifer (Lesne, 1895)
Bostrychus laminifer Lesne, 1895a: 174 [MNHN]
Distribution: Nt: AY (AYN) BV BZ (BZC BZE BZM BZS) EC PG SM UR

Bostrychopsis optata Lesne, 1938

Bostrychopsis optata Lesne, 1938a: 388 [BMNH]

Distribution: **E:** EH KY

Bostrychopsis parallela (Lesne, 1895)

Bostrychus parallelus Lesne, 1895a: 174 [ISNB MNHN]

= *Bostrychopsis affinis* Lesne, 1899a: 536 [MNHN]

Distribution: **A_i:** AS (ASN ASS) IA (SH); **E_i:** MB MT SO TZ ZA; **I:** BM CA IA (IAB IAC IAJ IAS) ID (IDB IDC IDE) LO MY (MYC) PH TH VT; **N_i:** US; **P:** AU_i CH (CE NO SE SW HAI TAI) FR_i GE_i ID (UP) JA YE_i (YEC)

Bostrychopsis peruana (Borowski et Węgrzynowicz, 2007)

Dominikia peruana Borowski et Węgrzynowicz, 2007a: 18 [ZMUB CASC]

= *Bostrichus eremita* Erichson, 1847: 87 [ZMUB CASC]

Distribution: **Nt:** AY CB PE PG UR

Bostrychopsis reichei (Marseul, 1867)

Apate reichei Marseul, 1867: 35 [MNHN]

Distribution: **E:** EH MB ML NI NX SG SO SU; **P:** AG EG LB SY TU

Bostrychopsis roonwali Rai, 1965

Bostrychopsis roonwali Rai, 1965: 576 [IFRI]

Distribution: **P:** ID (UP)

Bostrychopsis rostrifrons Lesne, 1923

Bostrychopsis rostrifrons Lesne, 1923a: 58 [MNHN]

Distribution: **E:** CG GX

Bostrychopsis scopula Lesne, 1923

Bostrychopsis scopula Lesne, 1923a: 57 [MNHN MRAC]

Distribution: **E:** KY

Bostrychopsis tabori Reichardt, 1962

Bostrychopsis tabori Reichardt, 1962a: 18 [MZSP]

Distribution: **Nt:** BZ (BZM)

Bostrychopsis tetraodon (Fairmaire, 1883)

Apate tetraodon Fairmaire, 1883c: 205 [MNHN]

Distribution: **E:** EH ER SU TZ; **N_i:** US (UST); **Nt:** TT; **P:** EG IT_i

Bostrychopsis tonsa (Imhoff, 1843)

Apate tonsa Imhoff, 1843: 177 [NHMB BMNH]

Distribution: E: AA BN CF CK CG CK EH EQ ER GH GO GX IV LI MZ NX SF SG SO SR SU TG TZ UG ZA; $\mathbf{Nt_i}$: BZ

Bostrychopsis trimorpha Lesne, 1899

Bostrychopsis trimorpha Lesne, 1899a: 550 [MNHN NHRS NHMW ZMUB]

Distribution: Nt: AY BV BZ CB PG VE

Bostrychopsis uncinata (Germar, 1824)

Apate uncinata Germar, 1824: 463 [ZMUB]

= *Amphicerus frontalis* Linell, 1899: 257 [USNM]

= *Apate affinis* Dejean, 1835: 308 [MNHN?] [**NN**]

= *Apate furcata* Perty, 1832: 83 [ZSMC]

= *Apate quadridentata* Dejean, 1835: 308 [MNHN?] [**NN**]

= *Apate serrata* Blanchard in Blanchard et Brullé, 1843: 204 [MNHN]

= *Bostrychopsis castelnaui* Bruch, 1915: 260 [MNHN] [**NN**]

= *Bostrychopsis gounellei* Bruch, 1915: 260 [MNHN] [**NN**]

= *Bostrychopsis orbignyi* Bruch, 1915: 260 [MNHN] [**NN**]

= *Bostrychopsis uncinata* morpha *castelnaui* Lesne, 1899a: 547 [MNHN] [**IN**]

= *Bostrychopsis uncinata* morpha *gounellei* Lesne, 1899a: 548 [MNHN] [**IN**]

= *Bostrychopsis uncinata* morpha *orbignyi* Lesne, 1899a: 547 [MNHN] [**IN**]

Distribution: Nt: AY (AYN) BV BZ (BZC BZM BZS) CB EC (GI) FG PE PG SM UR VE

Bostrychopsis valida Lesne, 1899

Bostrychopsis valida Lesne, 1899a: 544 [MNHN]

= *Bostrychopsis valida guyanica* Lesne, 1937e: 173 [BMNH]

Distribution: Nt: BV BZ (BZA BZE) GU; $\mathbf{P_i}$: GE

Bostrychopsis villosula Lesne, 1905

Bostrychopsis villosula Lesne, 1905b: 298 [MNHN NHMW SAMC]

= *Bostrichus cephalotes* Olivier, 1790a: 108 [**ND**]

Distribution: E: AA BD CG MZ KY RI RW SF TZ UG ZA ZI

Genus *Calophorus* Lesne, 1906

Calophorus Lesne, 1906a: 404

Type species: *Calophorus coriaceus* Lesne, 1906

Calophorus coriaceus Lesne, 1906

Calophorus coriaceus Lesne, 1906a: 405 [BMNH]

Distribution: A: AS (ASQ)

Calophorus sinoxylura Lesne, 1937

Calophorus sinoxylura Lesne, 1937e: 169 [MNHN]

= *Calophorus incisifrons* Lesne, 1937e: 171 [ANIC]

Distribution: **A:** AS (ASE ASN ASQ)

Genus *Dexicrates* Lesne, 1899

Dexicrates Lesne, 1899a: 455

Type species: *Bostrichus robustus* Blanchard, 1851

Dexicrates robustus (Blanchard, 1851)

Bostrichus robustus Blanchard, 1851: 433 [MNHN]

= *Apate curta* Dejean, 1835: 309 [MNHN?] [NN]

= *Dexicrates robustus argentinus* Lesne, 1911e: 345 [MNHN]

Distribution: **Nt:** AY (AYN) CL (CLN)

Genus *Dolichobostrychus* Lesne, 1899

Dolichobostrychus Lesne, 1899a: 583

Type species: *Bostrychus angustus* Steinheil, 1872

Dolichobostrychus angustus (Steinheil, 1872)

Bostrychus angustus Steinheil, 1872: 574 [MNHN]

= *Neoterius vitis* Mendes, 1932: 30 [IBSP DEES DEFS]

Distribution: **Nt:** AY (AYN AYS) BZ (BZC BZS) CB CL (CLN) PE PG VE

Dolichobostrychus fossulatus (Blanchard in Blanchard et Brullé, 1843)

Apate fossulata Blanchard in Blanchard et Brullé, 1843: 204 [MNHN]

Distribution: **Nt:** AY (AYN)

Dolichobostrychus gracilis (Lesne, 1899)

Neoterius gracilis Lesne, 1899a: 586 [MNHN]

Distribution: **Nt:** AY (AYN) BZ (BZC BZS)

Dolichobostrychus granulifrons (Lesne, 1895)

Bostrychus granulifrons Lesne, 1895a: 170 [MNHN ISNB]

Distribution: **Nt:** BZ (BZE BZM) VE

Dolichobostrychus yunnanus Lesne, 1913

Dolichobostrychus yunnanus Lesne, 1913a: 191 [MNHN]

Distribution: **P:** CH (SW) PA

†Genus *Discoclavata* Poinar, 2013

Discoclavata Poinar, 2013: 108

Type species: *Discoclavata dominicana* Poinar, 2013

†*Discoclavata dominicana* Poinar, 2013

Discoclavata dominicana Poinar, 2013: 108 [OSUO]

Distribution: Nt: DR (Dominican amber)

Genus *Heterobostrychus* Lesne, 1899

Heterobostrychus Lesne, 1899a: 554

Type species: *Bostrichus aequalis* Waterhouse, 1884

Heterobostrychus aequalis (Waterhouse, 1884)

Bostrichus aequalis Waterhouse, 1884: 215 [BMNH]

= *Bostrychus uncipennis* Lesne, 1895a: 173 [MNHN ISNB]

= *Rhizopertha papuensis* MacLeay, 1886: 154 [MAMU]

Distribution: A: AS (ASE) IA (IAN SH) MI NA$_i$ NZ$_i$ PW; **E$_i$:** CM CG GA GH MB SF ZA; **I:** AI BM CX IA (IAJ IAS) ID LO MY (MYC) PH TH VT; **N$_i$:** CN (CNM) US (USC USE USO UST); **Nt$_i$:** BI BR BZ CMS CU GL PA SM VE; **P:** BT CH (CE SE HAI SW TAI) FR$_i$ GB$_i$ GE$_i$ ID (HP UP) IR$_i$ IS$_i$ IT$_i$ JA MA$_i$ NL$_i$ NP PA PT$_i$ SP$_i$ SV$_i$

Heterobostrychus ambigenus Lesne, 1920

Heterobostrychus ambigenus Lesne, 1920: 295 [MNHN]

Distribution: P: CH (CE SW)

Heterobostrychus brunneus (Murray, 1867)

Bostrichus brunneus Murray, 1867: 92 [BMNH]

= *Bostrichus picipennis* Fåhraeus, 1871: 669 [NHRS]

= *Bostrychus brevicornis* Quedenfeldt, 1887: 326 [NN]

= *Bostrychus grayanus* Wollaston, 1867: 109 [BMNH]

Distribution: A$_i$: NZ; **E:** AA BD BF CG CK CV EH EQ ER GA GH GO GS GX IV KY MB MZ NI NX RW SF SG SO SU SYC TG TZ UG ZA ZB ZI; **N$_i$:** US (USD USG USL); **P$_i$:** BE CZ EG FR GB GE IT PL SP

Heterobostrychus hamatipennis (Lesne, 1895)

Bostrychus hamatipennis Lesne, 1895a: 173 [MNHN BMNH]

= *Apate niponensis* Lewis, 1896: 339 [SMFD BMNH?]

Distribution: A: PI; **E:** CM MB MU RI; **I:** AI BM CX IA (IAB IAJ) ID LO MY (MYC) PH TH VT; **N$_i$:** CN (CNM) US (USE); **P:** BE BT CH (CE HAI NE NO SE SW TAI) FR$_i$ GE$_i$ ID (HP SD) JA$_i$ NP OM PA SC

Heterobostrychus pileatus Lesne, 1899

Heterobostrychus pileatus Lesne, 1899a: 559 [MNHN BMNH]

Distribution: I: BM CA ID LO PH TH VT; **P:** CH (SW) NP

Heterobostrychus unicornis (Waterhouse, 1879)

Bostrichus unicornis Waterhouse, 1879a: 361 [BMNH]

Distribution: E: CM MB MZ; **I:** BM ID (IDC) VT; **P:** CH (SW) JA

Genus *Lichenophanes* Lesne, 1899

Lichenophanes Lesne, 1899a: 457

Type species: *Bostrichus tristis* Fåhraeus, 1871

Lichenophanes albicans Lesne, 1899

Lichenophanes albicans Lesne, 1899a: 491 [MNHN]

Distribution: Nt: AY (AYN) BZ (BZS)

Lichenophanes angustus (Casey, 1898)

Bostrichus angustus Casey, 1898: 72 [USNM]

= *Lichenophanes mutchleri* Belkin, 1940: 192

Distribution: N: CN (CNL) US (USL UST)

Note: Belkin (1940) designated a new name for *Lichenophanes angustus* (Casey, 1898). Steinhal (1872) described *Dolichobostrychus angustus* (Steinhal, 1872) as *Bostrychus* and not *Bostrichus*.

Lichenophanes arizonicus Fisher, 1950

Lichenophanes arizonicus Fisher, 1950: 77 [USNM NYBM]

Distribution: N: US (USO)

Lichenophanes armiger (LeConte, 1866)

Bostrichus armiger LeConte, 1866: 100 [MCZC]

Distribution: N: CN (CNL CNP) US (USD USE USG USL USO UST USU)

Lichenophanes bechyneorum Vrydagh, 1959

Lichenophanes bechyneorum Vrydagh, 1959c: 1 [ISNB NHMB CASC]

Distribution: Nt: AY (AYN)

Lichenophanes bedeli (Lesne, 1895)

Bostrychus bedeli Lesne, 1895a: 172 [MNHN]

Distribution: Nt: BZ (BZM BZS) CB

Lichenophanes bicornis (Weber, 1801)

Apate bicornis Weber, 1801: 91 [ZMUC?]

= *Apate deflexicornis* Sturm, 1843: 231 [NN]

Distribution: N: CN (CNL CNM) US (USD USE USG USL USO UST); **Nt_i:** PG

Lichenophanes californicus (Horn, 1878)

Bostrichus californicus Horn, 1878: 545 [PANS]

Distribution: N: US (USC)

Lichenophanes carinipennis (Lewis, 1896)

Apate carinipennis Lewis, 1896: 338 [BMNH]

= *Apate carinatus* Lewis, 1896: 339 [BMNH]

= *Bostrychus guttatus* Matsumura, 1915a: 129 [EIHU]

= *Bostrychus khmerensis* Lesne, 1896b: 511 [MNHN]

Distribution: I: AI BM CA CX MY (MYC) TH; **P:** CH (CE HAI SW TAI) JA SC

Lichenophanes caudatus (Lesne, 1895)

Bostrychus caudatus Lesne, 1895a: 172 [MNHN BMNH]

= *Lichenophanes fuliginosa* Lesne, 1899a: 485 [NN]

Distribution: E: CG CK EQ GH GO GS GX IV NX SG ZA; **Nt_i:** BZ

Lichenophanes collarti Vrydagh, 1959

Lichenophanes collarti Vrydagh, 1959c: 3 [ZMUB]

Distribution: E: TZ

Lichenophanes corticeus Lesne, 1908

Lichenophanes corticeus Lesne, 1908a: 34 [ZMUB]

Distribution: E: TZ

Lichenophanes egenus Lesne, 1923

Lichenophanes egenus Lesne, 1923a: 56 [MNHN MRAC MGDL]

= *Lichenophanes egenus* race *exciscus* Lesne, 1924: 123 [MNHN] [IN]

Distribution: E: CG CK EQ GO ZA

Lichenophanes egregius Lesne, 1934

Lichenophanes egregius Lesne, 1934c: 39 [MNHN]

Distribution: Nt: PE

Lichenophanes fasciatus (Lesne, 1895)

Bostrychus fasciatus Lesne, 1895a: 172 [MNHN]

Distribution: E_i: „West Africa“; **Nt:** BZ CB PG

Lichenophanes fascicularis (Fåhraeus, 1871)

Bostrichus fascicularis Fåhraeus, 1871: 670 [NHRS]

= *Apate morbillosa* Dejean, 1835: 309 [MNHN?] [NN]

= *Bostrychus morbillosus* Quedenfeldt, 1887: 325 [MNHN] [HN]

Distribution: E: AA CG CK CM EH EQ GO KY MAY MZ RW SF TZ UG ZA ZB

Lichenophanes funebris Lesne, 1938

Lichenophanes funebris Lesne, 1938c: 200 [BMNH]

Distribution: E: ZB

Lichenophanes indutus Lesne, 1935

Lichenophanes indutus Lesne, 1935a: 5 [MRAC ISNB]

= *Lichenophanes indutus* Lesne in Collart, 1934: 244 [NN]

Distribution: E: CG ZA

Lichenophanes iniquus (Lesne, 1895)

Bostrychus iniquus Lesne, 1895a: 171 [MNHN]

Distribution: E: BF CK GO GX SR TG ZA

Lichenophanes insignitus (Fairmaire, 1883)

Apate insignata Fairmaire, 1883a: 95 [MNHN]

Distribution: E: EH

Lichenophanes katanganus Lesne, 1935

Lichenophanes katanganus Lesne, 1935a: 7 [MRAC]

= *Lichenophanes katanganus* forma *major* Basilewsky, 1952: 105 [IN]

Distribution: E: ZA

Lichenophanes kuenckeli Lesne, 1895

Lichenophanes kuenckeli Lesne, 1895b: 178 [MNHN]

Distribution: E: MB

Note: Lesne (1895) described this species as „*künckeli*“ so the correct name is *Lichenophanes kuenckeli* (according to ICZN rules)

Lichenophanes marmoratus Lesne, 1908

Lichenophanes marmoratus Lesne, 1908b: 180 [MNHN]

= *Lichenophanes fascicularis* race *marmoratus* Lesne, 1899a: 477 [MNHN] [IN]

Distrbution: E: BN CG CK EQ NX SU ZA

Note: Lesne (1899) is commonly cited as the author of this species. This name is given here as a „race“ for the species *L. fascicularis* (Fåhraeus, 1871), the description that it is infrasubspecific name. Currently it is a nomen nudum, because there is no description and differential diagnosis (apart from two pictures). It was only mentioned as a valid species by Lesne (1908), including a short description and differential diagnosis.

Lichenophanes martini Lesne, 1899

Lichenophanes martini Lesne, 1899a: 501 [MNHN]

Distribution: E: MB

Lichenophanes numida Lesne, 1899

Lichenophanes numida Lesne, 1899a: 478 [MNHN]

Distribution: E: MT; **P:** AG IT (Sardinia I.) MO PT SP TU

Lichenophanes oberthuri Lesne, 1899

Lichenophanes oberthuri Lesne, 1899a: 478 [MNHN]

Distribution: E: CK GO NX TZ ZA ZB ZI

Lichenophanes penicillatus (Lesne, 1895)

Bostrychus penicillatus Lesne, 1895a: 171 [MNHN]

= *Bostrichus fasciculatus* Fall, 1909: 162 [MCZC]

Distribution: N: MX (MXC) US (USC); **Nt:** CC GT MX (MXS MXT)

Lichenophanes percristatus Lesne, 1924

Lichenophanes percristatus Lesne, 1924: 125 [MNHN]

Distribution: E: GO ZA

Lichenophanes perrieri Lesne, 1899

Lichenophanes perrieri Lesne, 1899a: 501 [MNHN]

Distribution: E: MB

Lichenophanes plicatus (Guérin-Méneville, 1844)

Bostrichus plicatus Guérin-Méneville, 1844: 185 [ISNB]

= *Apate inaequalis* Dejean, 1835: 309 [MNHN?] [NN]

= *Apate plicata* Sturm, 1843: 231 [unknown] [NN]

Distribution: Nt: AY (AYN) BZ (BZC BZM BZS) CB FG GT PG VE

Lichenophanes rutilans Reichardt, 1970

Lichenophanes rutilans Reichardt, 1970: 216 [ISNB MZSP CASC]

Distribution: Nt: EC (GI)

Lichenophanes spectabilis (Lesne, 1895)

Bostrychus spectabilis Lesne, 1895a: 173 [MNHN]

Distribution: N: US (USC); **Nt:** MX (MXM)

Lichenophanes szujeckii Węgrzynovicz et Borowski, 2015

Lichenophanes szujeckii Węgrzynovicz et Borowski, 2015a: 404 [DFPC]

Distribution: E: MZ ZB

Lichenophanes tristis (Fåhraeus, 1871)

Bostrichus tristis Fåhraeus, 1871: 669 [NHRS]

Distribution: E: SF

Lichenophanes truncaticollis (LeConte, 1866)

Bostrichus truncaticollis LeConte, 1866: 101 [MCZC]

Distribution: N: CN (CNL) US (USD USE USL USO UST)

Lichenophanes tuberosus Lesne, 1934

Lichenophanes tuberosus Lesne, 1934c: 40 [MNHN]

Distribution: Nt: MX (MXW)

Lichenophanes varius (Illiger, 1801)

Apate varia Illiger, 1801: 172 [ZMUB]

= *Apate gallica* Panzer, 1807: 17 [ZMUB]

= *Bostrichus dufourii* Latreille, 1807: 7 [MNHN?]

Distribution: P: AB AG AL AR AU BH BU CR CY CZ EG EN FR GE GG GR HU IN IT LB MC MD MO PL PT RO RU (CT ES ST) SK SL SP SY SZ TM TR TU UK YU

Lichenophanes verrucosus (Gorham, 1883)

Bostrychus verrucosus Gorham, 1883: 214 [BMNH]

Distribution: Nt: CC GT

Lichenophanes vespertinus Węgrzynovicz et Borowski, 2015

Lichenophanes vespertinus Węgrzynovicz et Borowski, 2015b: 574 [DFPC]

Distribution: E: GA

Lichenophanes weissi Lesne, 1908

Lichenophanes weissi Lesne, 1908b: 179 [MNHN]

= *Lichenophanes weissei* Mateu, 1974: 41 [LC]

Distribution: E: CG CK EQ GO GX IV ZA

Lichenophanes zanzibaricus Lesne, 1925

Lichenophanes zanzibaricus Lesne, 1925a: 28 [MNHN ZMUB ISNB]

Distribution: E: TZ TZ (Zanzibar)

Lichenophanes Lesne, 1899, other nomina nuda

Lichenophanes corticinus Burmeister in Heyne et Taschenberg, 1907: 197 [NN]

Genus *Micrapate* Casey, 1898

Micrapate Casey, 1898: 66

Type species: *Sinoxylon dinoderoides* Horn, 542

= *Bostrychulus* Lesne, 1899a: 591

Type species: *Sinoxylon dinoderoides* Horn, 1878

Micrapate albertiana Lesne, 1943

Micrapate kiangana albertiana Lesne, 1943: 33 [MRAC]

Distribution: E: ZA

Micrapate amplicollis (Lesne, 1899)

Bostrychulus amplicollis Lesne, 1899a: 615 [MNHN]

Distribution: Nt: PG

Micrapate atra (Lesne, 1899)

Bostrychulus ater Lesne, 1899a: 606 [MNHN]

Distribution: Nt: BZ (BZC BEZ)

Micrapate bicostula Lesne, 1906

Micrapate bicostula Lesne, 1906a: 406 [MNHN BMNH]

Distribution: Nt: BZ (BZC)

Micrapate bilobata Fisher, 1950

Micrapate bilobata Fisher, 1950: 94 [USNM]

Distribution: N: US (USO)

Micrapate brasiliensis (Lesne, 1899)

Bostrychulus brasiliensis Lesne, 1899a: 599 [MNHN NHMW]

= *Apate axillaris* Dejean, 1835: 309 [MNHN?] [NN]

Distribution: N_i: US (UST); **Nt:** BZ (BZC BZE BZS)

Micrapate brevipes (Lesne, 1899)

Bostrychulus brevipes Lesne, 1899a: 613 [MNHN]

Distribution: Nt: BZ (BZA)

Micrapate bruchi Lesne, 1931

Micrapate bruchi Lesne, 1931a: 99 [MNHN MACN]

Distribution: Nt: AY (AYN)

Micrapate brunnipes (Fabricius, 1801)

Apate brunnipes Fabricius, 1801: 383 [ZMUC]

= *Apate atratula* Lesne, 1899a: 604 [NN]

= *Xylopertha puncticollis* Kiesenwetter, 1877: 39 [ZSMC]

Distribution: Nt: „Antilles Is." BZ (BZC BZE) CB MX VE; **P$_i$:** FR GE IT

Micrapate catamarcana Lesne, 1931

Micrapate catamarcana Lesne, 1931a: 101 [MNHN MACN]

Distribution: Nt: AY

Micrapate cordobiana Lesne, 1931

Micrapate cordobiana Lesne, 1931a: 101 [MNHN MACN]

Distribution: Nt: AY

Micrapate cribripennis (Lesne, 1899)

Bostrychulus cribripennis Lesne, 1899a: 608 [MNHN]

Distribution: Nt: BZ (BZE)

Micrapate cristicauda Casey, 1898

Micrapate cristicauda Casey, 1898: 73 [USNM]

Distribution: N: US (USE USN USO UST)

Micrapate dinoderoides (Horn, 1878)

Sinoxylon dinoderoides Horn, 1878: 542 [PANS]

Distribution: N: MX (MXC) US (USO)

Micrapate discrepans Lesne, 1939

Micrapate discrepans Lesne, 1939b: 105 [MNHN]

Distribution: Nt: GT

Micrapate exigua (Lesne, 1899)

Bostrychulus exiguus Lesne, 1899a: 602 [MNHN]

Distribution: Nt: CB

Micrapate foraminata Lesne, 1906

Micrapate foraminata Lesne, 1906e: 276 [BMNH]

Distribution: Nt: CC PN

Micrapate fusca (Lesne, 1899)

Bostrychulus fuscus Lesne, 1899a: 603 [MNHN NHMW]

Distribution: Nt: CC CU MX (MXT) PN

Micrapate germaini (Lesne, 1899)

Bostrychulus germaini Lesne, 1899a: 609 [MNHN]
Distribution: **Nt:** BZ (BZM) PG

Micrapate guatemalensis Lesne, 1906

Micrapate guatemalensis Lesne, 1906e: 274 [BMNH MNHN]
Distribution: **Nt:** GT MX

Micapate horni (Lesne, 1899)

Bostrychulus horni Lesne, 1899a: 607 [MNHN]
Distribution: **Nt:** BZ (BZC)

Micrapate humeralis (Blanchard, 1851)

Bostrichus humeralis Blanchard, 1851: 434 [MNHN]
Distribution: **Nt:** CL (CLN)

Micrapate kiangana Lesne, 1935

Micrapate kiangana Lesne, 1935a: 12 [ZMUH]
Distribution: **E:** SSU TZ

Micrapate labialis Lesne, 1906

Micrapate labialis Lesne, 1906e: 278 [BMNH]
Distribution: **N_i:** US (USO UST); **Nt:** GT MX (MXT) PN

Micrapate leechi Vrydagh, 1960

Micrapate leechi Vrydagh, 1960a: 11 [CASC ISNB]
Distribution: **N:** US (USC)

Micrapate mexicana Fisher, 1950

Micrapate mexicana Fisher, 1950: 91 [USNM]
Distribution: **N:** US; **Nt:** MX

Micrapate neglecta Lesne, 1906

Micrapate neglecta Lesne, 1906a: 409 [BMNH]
Distribution: **E:** GA GX SR UG

Micrapate obesa (Lesne, 1899)

Bostrychulus obesa Lesne, 1899a: 614 [MNHN]
Distribution: **Nt:** BZ (BZE)

Micrapate pinguis Lesne, 1939

Micrapate pinguis Lesne, 1939b: 108 [BMNH]

Distribution: Nt: MX (MXT)

Micrapate puberula Lesne, 1906

Micrapate puberula Lesne, 1906a: 409 [MNHN]

Distribution: E: CD NX

Micrapate pupulus Lesne, 1906

Micrapate pupulus Lesne, 1906a: 407 [BMNH]

Distribution: Nt: BZ (BZA)

Micrapate quadraticollis (Lesne, 1899)

Bostrychulus quadraticollis Lesne, 1899a: 597 [ISNB]

Distribution: Nt: FG

Micrapate scabrata (Erichson, 1847)

Bostrichus scabratus Erichson, 1847: 87 [ZMUB]

= *Bostrichus vitis* Blanchard, 1851: 433 [MNHN ISNB]

Distribution: N_i: US (UST); **Nt:** BV CB CL (CLN) EC GI PE PN; P_i: AU

Micrapate scapularis (Gorham, 1883)

Xylopertha scapularis Gorham, 1883: 216 [BMNH MNHN]

Distribution: Nt: GT MX (MXT) PN

Micrapate schoutedeni Lesne, 1935

Micrapate schoutedeni Lesne, 1935a: 10 [MRAC]

= *Micrapate schoutedeni prolixula* Lesne, 1935a: 12 [MRAC MNHN]

Distribution: E: BD KY RW UG ZA

Micrapate sericeicollis Lesne, 1906

Micrapate sericeicollis Lesne, 1906e: 279 [BMNH ZMUH MNHN]

Distribution: Nt: BZ GT MX (MXT) UR

Micrapate simplicipennis (Lesne, 1895)

Xylopertha simplicipennis Lesne, 1895a: 177 [MNHN]

Distribution: I: BG BM IA (IAJ) ID (IDB) LO TH VT; **P:** CH (SW) NP ID (UP)

Micrapate straeleni Vrydagh, 1954

Micrapate straeleni Vrydagh, 1954: 32 [MRAC ISNB]

Distribution: E: CG MZ ZA

Micrapate unguiculata Lesne, 1906
Micrapate unguiculata Lesne, 1906e: 273 [BMNH]
Distribution: **Nt:** MX (MXT)

Micrapate wagneri Lesne, 1906
Micrapate wagneri Lesne, 1906c: 12 [MNHN]
Distribution: **Nt:** AY (AYN) BZ (BZM)

Micrapate xyloperthoides (Jacquelin do Val, 1859)
Apate xyloperthoides Jacquelin do Val, 1859: 40 [ISNB]
= *Apate phalaridis* Lesne, 1899a: 610 [**NN**]
Distribution: **N_i:** US (UST); **P:** AG FR GE IT MO PT SP TU

Genus *Neoterius* Lesne, 1899
Neoterius Lesne, 1899a: 581
Type species: *Bostrichus mystax* Blanchard, 1851

Neoterius fairmairei (Lesne, 1895)
Bostrychus fairmairei Lesne, 1895a: 171 [MNHN]
Distribution: **Nt:** CL (CLN) MX PE

Neoterius mystax (Blanchard, 1851)
Bostrichus mystax Blanchard, 1851: 432 [MNHN]
Distribution: **N:** BZ CL (CLN) PE

Neoterius pulvinatus (Blanchard, 1851)
Bostrichus pulvinatus Blanchard, 1851: 431 [ISNB]
Distribution: **Nt:** BZ CL (CLN)

Genus *Parabostrychus* Lesne, 1899
Parabostrychus Lesne, 1899a: 590
Type species: *Bostrychus elongatus* Lesne, 1895

Parabostrychus acuticollis Lesne, 1913
Parabostrychus acuticollis Lesne, 1913a: 192 [MNHN NZSI]
Distribution: **I:** ID (IDE IDW) TH VT; **P:** CH (CE NO SE SW TAI) NP ID (UP)

Parabostrychus elongatus (Lesne, 1895)
Bostrychus elongatus Lesne, 1895a: 170 [MNHN]
Distribution: **I:** VT; **P:** CH (SW) ID (UP)

Genus *Sinoxylodes* Lesne, 1899

Sinoxylodes Lesne, 1899a: 617

Type species: *Bostrichus curtulus* Erichson, 1847

Sinoxylodes curtulus (Erichson, 1847)

Bostrichus curtulus Erichson, 1847: 87 [ZMUB]

= *Sinoxylon championi* Gorham, 1883:214 [BMNH]

Distribution: **Nt:** AY (AYN) BV BZ (BZM) GT PE PG

Tribe Sinoxylini Marseul, 1857

Sinoxylidae Marseul, 1857: 107

Type genus: *Sinoxylon* Duftschmid, 1825

= Sinoxylini Liu & Schönitzer, 2011 (sic!) = Sinoxylini Liu, Leavengood et Bernal, 2022: 25 [**HN**]

= Sinoxyloninae Lesne, 1899a: 439 [**HN**]

Type genus: *Sinoxylon* Duftschmid, 1825

Genus *Calodectes* Lesne, 1906

Calodectes Lesne, 1906b: 453

Type species: *Calodectes laniger* Lesne, 1906

Calodectes laniger Lesne, 1906

Calodectes laniger Lesne, 1906b: 454 [SAMC]

Distribution: **E:** SF

Genus *Calodrypta* Lesne, 1906

Calodrypta Lesne, 1906b: 455

Type species: *Calodrypta exarmata* Lesne, 1906

Calodrypta exarmata Lesne, 1906

Calodrypta exarmata Lesne, 1906b: 455 [MNHN BMNH]

Distribution: **E:** CG MZ RW SF TG TZ UG ZA ZB

Genus *Calopertha* Lesne, 1906

Calopertha Lesne, 1906b: 456

Type species: *Sinoxylon subretusum* Ancey, 1881

Calopertha costatipennis Lesne, 1906

Calopertha costatipennis Lesne, 1906b: 458 [MNHN MCSN MCZR]

Distribution: **E:** DJ EH SO; **P:** YE (YEC)

Calopertha kalaharensis Lesne, 1906

Calopertha kalaharensis Lesne, 1906b: 459 [MNHN ZMUB]

Distribution: E: NB SF

Calopertha subretusa (Ancey, 1881)

Sinoxylon subretusum Ancey, 1881: 509 [ZMUH or MNHN]

Distribution: E: CD EH ER KY MT NI NX SG SO SR SU TZ; I: ID; P: BT EG MO OM SA YE (YEC)

Calopertha truncatula (Ancey, 1881)

Sinoxylon truncatulum Ancey, 1881: 509 [ZMUH or MNHN]

Distribution: E: CD CK DJ EH ER KY MT NI NX SG SO SU TZ; I: ID (IDP); P: AE AF BT EG IN IQ IS MO OM PA SA YE (YEC)

Genus *Sinocalon* Lesne, 1906

Sinocalon Lesne, 1906b: 447

Type species: *Bostrychus vestitus* Lesne, 1895

Sinocalon pilosulum Lesne, 1906

Sinocalon pilosulum Lesne, 1906b: 452 [MNHN NHMW BMNH]

Distribution: Nt: AY (AYN AYS)

Sinocalon reticulatum Lesne, 1906

Sinocalon reticulatum Lesne, 1906b 451 [ZMUB MNHN]

Distribution: Nt: AY (AYN AYS)

Sinocalon vestitum (Lesne, 1895)

Bostrychus vestitus Lesne, 1895a: 175 [MNHN ISNB]

Distribution: Nt: AY (AYN AYS) BV

Genus *Sinoxylon* Duftschmid, 1825

Sinoxylon Duftschmid, 1825: 85

Type species: *Bostrichus sexdentatus* Olivier, 1790

= *Apatodes* Blackburn, 1889: 1429

Type species: *Apatodes macleayi* Blackburn, 1889

= *Trypocladus* Guérin-Méneville, 1845: 17

Type species: *Bostrichus sexdentatus* Olivier, 1790

Sinoxylon anale Lesne, 1897

Sinoxylon anale Lesne, 1897b: 21 [MNHN BMNH ISNG]

= *Apatodes macleayi* Blackburn, 1889: 1429 [SAMA]

= *Sinoxylon geminatum* Schilsky, 1899: 80 [ZMUB]

Distribution: A_i: AS (ASN ASS) IA (SH) NZ PW „Mellvile I.“ HI; E_i: CG GX ZA; I: BM CA CX IA (IAC IAJ IAS) ID (IDC) MY (MYC) PH RI TH VT; N_i: CN (CNM) US (USC USD USE USL UST); Nt_i: AY BZ (BZC BZS) PG UR VE; P_i: AU BE CH (CE NO NW SE SW TAI) FI FR GB GE GR ID (HP UP) IN IQ IS IT IQ JA NP PA PL SP UK

Sinoxylon angolense Lesne, 1906

Sinoxylon angolense Lesne, 1906b: 507 [MNHN]

Distribution: E: AA GO ZA

Sinoxylon atratum Lesne, 1897

Sinoxylon atratum Lesne, 1897b: 20 [MNHN IFRI]

= *Sinoxylon atratum kohlarianum* Lesne, 1906b: 540 [MNHN]

Distribution: I: ID (IDC IDW) TH; P: CH (NW) ID (HP)

Sinoxylon beesoni Lesne, 1931

Sinoxylon beesoni Lesne, 1931a: 102 [MNHN IFRI]

Distribution: I: BM

Sinoxylon bellicosum Lesne, 1906

Sinoxylon bellicosum Lesne, 1906b: 497 [BMNH SAMC MNHN]

Distribution: E: MZ SF TZ ZA

Sinoxylon birmanum Lesne, 1906

Sinoxylon birmanum Lesne, 1906b: 540 [NHMW]

Distribution: I: BM; P_i: GE

Sinoxylon brazzai Lesne, 1895

Sinoxylon brazzai Lesne, 1895a: 177 [MNHN]

Distribution: E: BN CG CK GH GO GX IV NX SR TG TZ UG ZA

Sinoxylon bufo Lesne, 1906

Sinoxylon bufo Lesne, 1906b: 482 [MNHN ZMUB]

Distribution: I: IA (IAB IAJ) PH

Sinoxylon cafrum Lesne, 1905

Sinoxylon cafrum Lesne, 1905c: 276 [MNHN ZMUH BMNH OXUM]

Distribution: E: MZ NB SF TZ

Sinoxylon capillatum Lesne, 1895

Sinoxylon capillatum Lesne, 1895a: 175 [MNHN]

Distribution: P: ID (KA UP) PA

Sinoxylon ceratoniae (Linnaeus, 1758)
Scarabeus ceratoniae Linnaeus, 1758: 353 [unknown]
= *Apate diaspis* Fairmaire in Gestro, 1895: 357 [NN]
= *Sinoxylon bicuspidatum* Ancey, 1879: 139 [MNHN]
Distribution: **E:** CD CG CK CV EH ER GX KY MB MT MZ NB NI NX SF SG SO SSU TZ ZA; **Nt_i:** EC; **P:** AE AF AG EG GE_i IQ IS LB MO OM SA SV_i TU YE (YEC)

Sinoxylon circuitum Lesne, 1897
Sinoxylon circuitum Lesne, 1897b: 22 [MNHN]
Distribution: **I:** ID (IDC)

Sinoxylon crassum Lesne, 1897
Sinoxylon crassum Lesne, 1897b: 21 [MNHN BMNH ISNB]
= *Sinoxylon coptura* Lesne, 1897b: 21 [NN]
= *Sinoxylon crassum dekkanense* Lesne, 1906b: 538 [ISNB MNHN ZMUB BMNH]
Distribution: **A_i:** AS; **E_i:** TZ; **I:** BM CA CX ID (IDC) LO MY (MYC) PH TH VT; **P:** AE CH (TAI) GE_i ID (HP UP) IN NP OM_i PL_i

Sinoxylon cucumella Lesne, 1906
Sinoxylon cucumella Lesne, 1906b: 544 [MNHN]
Distribution: **I:** ID (IDE) LO TH VT; **P:** BT CH (SW) ID (SD) NP

Sinoxylon cuneolus Lesne, 1906
Sinoxylon cuneolus Lesne, 1906b: 508 [BMNH ZMUB]
Distribution: **E:** SF

Sinoxylon dichroum Lesne, 1906
Sinoxylon dichroum Lesne, 1906b: 508 [MNHN]
Distribution: **I:** BM ID (IDE) VT; **P:** CH (SW) ID (UP) NP

Sinoxylon divaricatum Lesne, 1906
Sinoxylon divaricatum Lesne, 1906b: 495 [MNHN]
Distribution: **E:** KY SF SO TZ

Sinoxylon epipleurale Lesne, 1906
Sinoxylon epipleurale Lesne, 1906b: 496 [MNHN BMNH]
Distribution: **E:** MZ NB TZ ZA ZB ZI

Sinoxylon erasicauda Lesne, 1906
Sinoxylon erasicauda Lesne, 1906b: 502 [ISNB]
Distribution: **E:** TZ ZB

Sinoxylon eucerum Lesne, 1932

Sinoxylon eucerum Lesne, 1932b: 657 [MNHN]

Distribution: P: CH (CE NO)

Sinoxylon flabrarius Lesne, 1906

Sinoxylon flabrarius Lesne, 1906b: 543 [MNHN BMHN]

Distribution: I: BM ID LO TH VT; **P:** BT CH (HKG SW) GB_i NP

Sinoxylon fuscovestitum Lesne, 1919

Sinoxylon fuscovestitum Lesne, 1919: 465 [MNHN]

Distribution: I: BM LO TH VT; **P:** BT CH (SW) ID (SD) NP SA_i

Sinoxylon indicum Lesne, 1897

Sinoxylon indicum Lesne, 1897b: 22 [MNHN ISNB BMNH IFRI]

Distribution: I: BM ID (IDC IDE „Deccan Plateau" „Hindustan"); N_i: US (USE); **P:** IN IT_i NP PA

Sinoxylon japonicum Lesne, 1895

Sinoxylon japonicum Lesne, 1895a: 175 [MNHN]

Distribution: N_i: US; **P:** CH (CE NO NW SE SW TAI) JA SC

Sinoxylon lesnei Vrydagh, 1955

Sinoxylon lesnei Vrydagh, 1955b: 7 [ZMUB MNHN]

Distribution: E: GA GS GX TG

Sinoxylon luzonicum Lesne, 1932

Sinoxylon luzonicum Lesne, 1932a: 393 [MNHN UPPC]

Distribution: I: PH

Sinoxylon lycturum Lesne, 1936

Sinoxylon lycturum Lesne, 1936a: 136 [IFRI]

Distribution: I: BM

Sinoxylon malaccanum Lesne, 1930

Sinoxylon malaccanum Lesne, 1930: 103 [MNHN BMNH]

Distribution: I: MY

Sinoxylon mangiferae Chûjô, 1936

Sinoxylon mangiferae Chûjô, 1936: 407 [NTUC]

Distribution: I: ID (IDE) LO TH; **P:** CH (CE HAI SE SW TAI) ID (SD) NP

Sinoxylon marseuli Lesne, 1895
Sinoxylon marseuli Lesne, 1895a: 177 [MNHN IFRI]
= *Sinoxylon marseuli convexicauda* Lesne, 1932b: 657 [MNHN IFRI]
Distribution: I: IA (IAC IAJ IAS) ID (IDE) MY (MYC) VT

Sinoxylon oleare Lesne, 1932
Sinoxylon oleare Lesne, 1932b: 655 [IFRI MNHN]
Distribution: P: ID (HP UP)

Sinoxylon pachyodon Lesne, 1906
Sinoxylon pachyodon Lesne, 1906b: 486 [MCSN]
Distribution: I: BM ID TH

Sinoxylon parviclava Lesne, 1918
Sinoxylon parviclava Lesne, 1918: 490 [BMNH MNHN]
Distribution: I: BM CA IA (IAC) PH TH VT

Sinoxylon perforans (Schrank von Paula, 1789)
Bostrichus perforans Schrank von Paula, 1789: 64 [NHMW?]
= *Bostrichus bidentatus* Rossi, 1790: 525 [ZMUB?]
= *Bostrichus bispinosus* Olivier, 1790a: 110 [BMNH]
Distribution: P: AB AL AR AU BH BU CR CY CZ FR GE GG GR HU IN IS IT MC MD MO PL PT RO RU (ST) SK SL SP SY SZ TM TR UK YU

Sinoxylon philippinense Lesne, 1938
Sinoxylon philippinense Lesne, 1938d: 404 [BMNH]
Distribution: I: PH

Sinoxylon pubens Lesne, 1906
Sinoxylon pubens Lesne, 1906b: 511 [MNHN]
Distribution: I: ID (IDC); **P:** CH (SW)

Sinoxylon pugnax Lesne, 1904
Sinoxylon pugnax Lesne, 1904: 159 [NHMW NZSI MNHN]
Distribution: P: AF ID (KA) IN OM PA

Sinoxylon pygmaeum Lesne, 1897
Sinoxylon pygmaeum Lesne, 1897b: 20 [MNHN NZSI BMNH]
= *Sinoxylon pygmaeum annamitum* Lesne, 1938d: 404 [MNHN NZSI BMNH]
Distribution: I: BM ID (IDC IDE IDP) TH VT; **P:** CH (NW TAI) NP

Sinoxylon rejectum (Hope, 1845)

Apate rejecta Hope, 1845: 16 [BMNH?]

Distribution: **P:** CH (CE SE)

Sinoxylon ruficorne Fåhraeus, 1871

Sinoxylon ruficorne Fåhraeus, 1871: 665 [NHRS]

= *Sinoxylon ruficorne guineense* Lesne, 1906f: 413 [MCSN MNHN ZMUH]

Distribution: **E:** AA BD BF BN EH CD CG CK GA GS GX KY MB MT MZ NB NX RW SF SG SO STI SU TG TZ ZA; **N_i:** US; **Nt_i:** BZ; **P_i:** IT

Sinoxylon rufobasale Fairmaire, 1888

Sinoxylon rufobasale Fairmaire, 1888: 179 [MNHN]

= *Sinoxylon doliolum* Lesne, 1905c: 275 [ZMUH NHMW]

Distribution: **E:** CG GA KY MZ NB RW SF TZ UG ZA ZB; **P:** YE (YEC)

Sinoxylon rugicauda Lesne, 1929

Sinoxylon rugicauda Lesne, 1929a: 459 [ZRC]

Distribution: **I:** MY

Sinoxylon senegalense Karsch, 1881

Sinoxylon senegalense Karsch, 1881: 42 [ZMUB]

= *Apate dentifrons* Sturm, 1843: 231 [unknown] [**NN**]

= *Apate senegalensis* Dejean, 1835: 34 [MNHN?] [**NN**]

= *Apate senegalense dentifrons* Karsch, 1881: 42 [**NN**]

= *Sinoxylon coronatum* Zoufal, 1894: 38 [unknown]

= *Sinoxylon gloriosus* Rai et Chatterjee, 1963a: 15 [IFRI]

= *Sinoxylon senegalense vermiculatum* Lesne, 1906b: 501 [MNHN]

Distribution: **E:** BF CD CK CV EH ER EQ GA ML MT NI NX SF SG SR SO SSU SU TZ UG; **I:** ID; **Nt_i:** BZ; **P:** AG EG GE_i ID (UP) LB MO SA UK_i YE (YEC)

Sinoxylon sexdentatum (Olivier, 1790)

Bostrichus sexdentatus Olivier, 1790a: 110 [MNHN]

= *Apate bidens* Fabricius, 1798: 157 [ZMUC]

= *Apate muricata* Dejean, 1821: 101[MNHN?] [**NN HN**]

= *Apate muricata* Dejean, 1821: 101 [MNHN?] [**NN HN**]

= *Apate muricata* Fabricius, 1775: 54 [ZMUC?]

= *Apate muricata* Fabricius in Dejen, 1833: 309 [MNHN?] [**NN HN**]

= *Bostrichus chalcographus* Panzer, 1794a: 4 [ZMUB]

= *Dermestes muricatus* Linnaeus, 1767: 562 [unknown] [**NO**]

Distribution: **N_i:** US (UST); **P:** AB AG AL AU BH BU CH (SW) CR CY EG FR GE GR HU IN IS IT LB MO PT RU (CT ST) SL SP SY TR TU UK

Sinoxylon succisum Lesne, 1895

Sinoxylon succisum Lesne, 1895a: 176 [MNHN]

Distribution: **E:** BN GA GX ML MZ NX SG; **P:** AG MO TU

Sinoxylon sudanicum Lesne, 1895

Sinoxylon sudanicum Lesne, 1895a: 176 [MNHN]

Distribution: **E:** SG SU MB; **I:** BG CA ID (IDC); **N_i:** US (USE); **P:** CH (NW SW) EG GE_i IN IT_i OM PA YE (YEC)

Sinoxylon tignarium Lesne, 1902

Sinoxylon tignarium Lesne, 1902a: 116 [MNHN]

Distribution: **I:** ID (IDE) TH VT; **P:** CH (SW) ID (SD) NP

Sinoxylon transvaalense spathiferum Lesne, 1906

Sinoxylon transvaalense spathiferum Lesne, 1906b: 533 [MNHN MNMS]

Distribution: **E:** CD CG EH ER GA GS ML RW SG SU NX ZA

Sinoxylon transvaalense transvaalense Lesne, 1895

Sinoxylon transvaalense Lesne, 1895a: 176 [MNHN]

Distribution: **E:** AA BW MZ SF TZ UG ZA ZB

Sinoxylon unidentatum (Fabricius, 1801)

Sinodendron unidentatum Fabricius, 1801: 377 [ZMUC]

= *Sinoxylon conigerum* Gerstäcker, 1855: 268 [ZMUB]

Distribution: **A_i:** HI; **E:** CG GA KY MB MU MZ RI SO TZ ZA; **I:** CX IA (IAJ) ID PH TH; **N_i:** US (USE UST); **Nt_i:** AY BR BZ (BZC BZM) CB GI HA PG VE; **P_i:** CH FR GB GE GR CH (TAI) IT MA PL RU SP UK YE (YEC)

Sinoxylon verrugerum Lesne, 1906

Sinoxylon transvaalense verrugerum Lesne, 1906b: 534 [MNHN ISNB]

Distribution: **E:** ML NX TG ZI

Sinoxylon villosum Lesne, 1895

Sinoxylon villosum Lesne, 1895a: 176 [MNHN]

Distribution: **E:** SF

Genus *Xyloperthodes* Lesne, 1906

Xyloperthodes Lesne, 1906b: 545

Type species: *Sinoxylon nitidipenne* Murray, 1867

Xyloperthodes abruptus (Lesne, 1906)

Sinoxylon abruptus Lesne, 1906f: 414 [MNHN ZMUB]

Distribution: E: EQ GS

Xyloperthodes baudouini Vrydagh, 1956

Xyloperthodes baudouini Vrydagh, 1956d: 258 [MRAC]

Distribution: E: ZA

Xyloperthodes castaneipennis (Fåhraeus, 1871)

Xyloperthodes castaneipennis Fåhraeus, 1871: 666 [NHRS]

Distribution: E: IV KY MZ SF SU TZ ZB

Xyloperthodes clavula Lesne, 1906

Xyloperthodes clavula Lesne, 1906b: 553 [MNHN ZMUB]

Distribution: E: KY SF TZ TZ UG

Xyloperthodes collarti Lesne, 1935

Xyloperthodes collarti Lesne, 1935a: 13 [MRAC ISNB]

= *Xyloperthodes collarti* Lesne in Collart, 1934: 246 [NN]

Distribution: E: ZA

Xyloperthodes discedens Lesne, 1906

Xyloperthodes discedens Lesne, 1906a: 559 [MNHN]

Distribution: E: CG CK GO IV SR STI ZA

Xyloperthodes discicollis (Fairmaire, 1893)

Sinoxylon discicollis Fairmaire, 1893a: 27 [MNHN]

Distribution: E: EH ER SSU

Xyloperthodes evops Lesne, 1906

Xyloperthodes evops Lesne, 1906a: 556 [MNHN BMNH SAMC ZMUB]

Distribution: E: AA BD KY SF TZ ZA ZB ZI

Xyloperthodes granulatus Lesne, 1906

Xyloperthodes granulatus Lesne, 1906a: 558 [MNHN]

= *Xyloperthodes granulatus sianakensis* Lesne, 1906: 558 [MNHN]

Distribution: E: MB

Xyloperthodes houssiaui Vrydagh, 1955

Xyloperthodes houssiaui Vrydagh, 1955d: 263 [ISNB]

Distribution: E: ZA

Xyloperthodes hova Lesne, 1906

Xyloperthodes hova Lesne, 1906a: 552 [MNHN]

Distribution: E: MB

Xyloperthodes incertus Lesne, 1906

Xyloperthodes incertus Lesne, 1906a: 554 [BMNH SAMC MNHN MCSN]

Distribution: E: ER KY MZ SF TZ ZA ZB

Xyloperthodes nasifer Lesne, 1906

Xyloperthodes nasifer Lesne, 1906a: 557 [MNHN BMNH ZMUB]

Distribution: E: MB RI

Xyloperthodes nitidipennis (Murray, 1867)

Sinoxylon nitidipenne Murray, 1867: 94 [BMNH]

= *Apate spadicea* Dejean, 1835: 309 [MNHN?] [**NN**]

= *Xylopertha plagiata* Fåhraeus, 1871: 666 [NHRS]

= *Xylopertha polita* Quedenfeldt, 1887: 327 [ZMUB]

= *Xyloperthodes nitidipennis plagatus* Vrydagh, 1955a: 13 [**LC**]

Distribution: E: AA BD CG CK EH EQ GA GH GO GS GX IV KY LI MZ NX RW SF SG SR TG TZ UG ZA; **P:** GB_i GE_i MO

Xyloperthodes orthogonius Lesne, 1906

Xyloperthodes orthogonius Lesne, 1906a: 559 [MNHN]

Distribution: E: CG GA GX IV RW UG ZA ZI

Xyloperthodes pollicifer Vrydagh, 1955

Xyloperthodes pollicifer Vrydagh, 1955d: 265 [ISNB]

Distribution: E: ZA

Xyloperthodes schedli Vrydagh, 1959

Xyloperthodes schedli Vrydagh, 1959c: 8 [OLML ISNB]

Distribution: E: ZA

Tribe Xyloperthini Lesne, 1921

Xyloperthini Lesne, 1921c: 288

Type genus: *Xylopertha* Guérin-Méneville, 1845

Genus *Amintinus* Anonymous, 1939

Amintinus Anonymous, 1939: 241

Type species: *Amintinus aethiopicus* Lesne, 1938

= *Amintinus* Lesne, 1938a: 390 [**NN**]

= *Amintinus* Borowski et Węgrzynowicz, 2007a: 123

Type species: *Amintinus aethiopicus* Lesne, 1938

= *Amintinus* Borowski, 2020: 34 [NN]

Note: According to rules of ICZN an anonymous authorship before year 1951 is available.

Amintinus aethiopicus (Lesne, 1938)

Amintinus aethiopicus Lesne, 1938a: 391 [BMNH]

Distribution: E: EH

Amintinus gardneri Vrydagh, 1959

Amintinus gardneri Vrydagh, 1959c: 12 [ISNB]

Distribution: E: UG

Amintinus lootensi Damoiseau, 1968

Amintinus lootensi Damoiseau, 1968a: 310 [MRAC HNHM]

Distribution: E: CG ZA

Amintinus minutissimus Damoiseau, 1968

Amintinus minutissimus Damoiseau, 1968c: 4 [HNHM ISNB]

Distribution: E: CG ZA

Amintinus ruwenzorius Vrydagh, 1955

Amintinus ruwenzorius Vrydagh, 1955d: 261 [MRAC ISNB]

Distribution: E: ZA

Amintinus sakalavus Lesne, 1939

Amintinus sakalavus Lesne, 1939a: 36 [MNHN]

Distribution: E: MB

Amintinus subtilis Lesne, 1939

Amintinus subtilis Lesne, 1939a: 36 [MNHN BMNH]

Distribution: E: KY

Amintinus tenuis Lesne, 1938

Amintinus tenuis Lesne, 1938a: 393 [MNHN]

Distribution: E: KY UG ZA; **P:** ID (SD)

Genus *Calonistes* Lesne, 1936

Calonistes Lesne, 1936a: 137

Type species: *Calonistes antennalis* Lesne, 1936a

Calonistes antennalis Lesne, 1936
Calonistes antennalis Lesne, 1936a: 138 [MNHN BMNH]
Distribution: I: MY (MYC) TH

Calonistes vittatus Zhang, Meng et Beaver, 2022
Calonistes vittatus Zhang, Meng et Beaver, 2022: 528 [IZAS]
Distribution: P: CH (SW)

Genus *Ctenobostrychus* Reichardt, 1962
Ctenobostrychus Reichardt, 1962c: 173
Type species: *Ctenobostrychus alverneri* Reichardt, 1962

Ctenobostrychus alverneri Reichardt, 1962c
Ctenobostrychus alverneri Reichardt, 1962c: 174 [MZSP]
Distribution: Nt: BZ (BZM)

Genus *Dendrobiella* Casey, 1898
Dendrobiella Casey, 1898: 66
Type species: *Sinoxylon sericans* LeConte, 1858

Dendrobiella aspera (LeConte, 1858)
Sinoxylon asperum LeConte, 1858: 73 [MCZC]
Distribution: N: MX (MXC) US (USC USO USU); **Nt:** DR GT

Dendrobiella isthmicola Lesne, 1933
Dendrobiella isthmicola Lesne, 1933b: 238 [MNHN]
Distribution: Nt: CB CC MX (MXT) VE

Dendrobiella leechi Vrydagh, 1960
Dendrobiella leechi Vrydagh, 1960a: 15 [BMNH CASC]
Distribution: N: US (USC); **Nt:** GT MX PE

Dendrobiella sericans (LeConte, 1858)
Sinoxylon sericans LeConte, 1858: 73 [MCZC]
= *Apate mexicana* Dejean, 1835: 309 [MNHN?] [NN]
= *Dendrobiella pubescens* Casey, 1898: 67 [USNM]
Distribution: N: MX (MXC) US (USC USO); **Nt:** GT HO JC MX (MXM MXS MXT MXW) PN

Dendrobiella sericea (Mulsant et Wachanru, 1852)
Xylopertha sericea Mulsant et Wachanru, 1852: 14 [MNHN?]
= *Apate puberula* Dejean, 1835: 309 [MNHN?] [NN]
= *Dendrobiella sublaevis* Casey, 1898: 68 [USNM]

Distribution: **N:** CB CC CU DR GT HA JC MX (MXW)

Note: Mulsant & Wachanru (1852) described this species from Caramanie (south Anatolia, Turkey, Palaearctic Region).

Genus *Enneadesmus* Mulsant, 1851

Enneadesmus Mulsant, 1851: 208

Type species: *Bostrichus trispinosus* Olivier, 1790

Enneadesmus auricomus (Reitter, 1898)

Xylopertha auricoma Reitter, 1898: 13 [HNHM]

= *Bostrichus bidentatus* Olivier, 1790b: 16 [MNHN]

Distribution: **P:** IN SY TM UZ

Enneadesmus bigranulum Lesne, 1901

Enneadesmus bigranulum Lesne, 1901a: 603 [MNHN]

= *Enneadesmus brigranulatum* Schedl, 1964: 282 [LC]

Distribution: **E:** MB TZ

Enneadesmus crassispina Lesne, 1936

Enneadesmus crassispina Lesne, 1936b: 56 [MNHN ZMUB]

Distribution: **E:** MZ TZ ZB

Enneadesmus decorsei Lesne, 1901

Enneadesmus decorsei Lesne, 1901a: 604 [MNHN]

Distribution: **E:** MB

Enneadesmus evacanthus Lesne, 1901

Enneadesmus evacanthus Lesne, 1901a: 608 [MNHN MCSN]

Distribution: **E:** EH ER KY SO TZ ZB

Enneadesmus forficula forficula (Fairmaire, 1883)

Apate forficula Fairmaire, 1883a: 95 [MNHN ISNB]

= *Apate spinipennis* Dejean, 1835: 309 [MNHN?] [NN]

= *Enneadesmus forficula* race *capensis* Lesne, 1901a: 607 [NHRS ZMUB MNHN SAMC] [IN]

= *Enneadesmus forficula capensis* Vrydagh, 1961a: 15

= *Xylopertha auberti* Chobaut, 1897: 170 [MNHN]

Distribution: **E:** CD CK CV DJ EH ER KY ML MT MZ NI NX SF SG SSU SO SU TZ; **P:** AG EG GR ID (IDP) IN IS JO LB MO OM PA SA TU YE (YEC)

Enneadesmus forficula socotrensis Borowski in Borowski et Sławski, 2017

Enneadesmus forficula socotrensis Borowski in Borowski et Sławski, 2017: 108 [NMPC]

Distribution: **P:** YE (YES)

Enneadesmus mariae Lesne, 1936

Enneadesmus mariae Lesne, 1936b: 58 [MNHN BMNH SAMC]

Distribution: E: LH MZ SF ZA

Enneadesmus masculinus Lesne, 1936

Enneadesmus masculinus Lesne, 1936b: 57 [MNHN ZMUB]

Distribution: E: BW GA GX MZ NB SF TZ ZI

Enneadesmus nigritulus Lesne, 1937

Enneadesmus forficula nigritulus Lesne, 1937d: 86 [MNHN ZMUB MRAC]

Distribution: E: KY SO SU TZ UG

Enneadesmus obtusidentatus obtusidentatus (Lesne, 1899)

Xylopertha obtusidentata Lesne, 1899b: 228 [MNHN MCSN]

= *Enneadesmus obtusedentatus* Lesne, 1924: 211 [LC]

Distribution: E: DJ EH ER SO SU; **P:** EG GE$_{i}$ IN IQ OM SA SY YE (YEC)

Enneadesmus obtusidentatus obscurior Borowski in Borowski et Sławski, 2017

Enneadesmus obtusidentatus obscurior Borowski in Borowski et Sławski, 2017: 104 [NMPC]

Distribution: P: YE (YES)

Enneadesmus scopini Fursov, 1936

Enneadesmus scopini Fursov, 1936: 349 [ZMUM]

= *Enneadesmus scopini minor* Fursov, 1936: 349 [ZMUM]

Distribution: P: UZ

Enneadesmus sculptifrons Lesne, 1905

Enneadesmus sculptifrons Lesne, 1905c: 275 [ZMUH]

Distribution: E: SF

Enneadesmus trispinosus (Olivier, 1790)

Bostrichus trispinosus Olivier, 1790b: 16 [MNHN?]

= *Xylopertha lesnei* Chobaut, 1898: 81 [MNHN?]

Distribution: P: AG CI CR EG RU (FE) FR IN IQ IR IS IT LB MO OM PT SP TU

Genus *Gracilenta* Zhang, Meng et Beaver, 2022

Gracilenta Zhang, Meng et Beaver, 2022: 530

Type species: *Gracilenta yingjiangensis* Zhang, Meng et Beaver, 2022

Gracilenta yingjiangensis Zhang, Meng et Beaver, 2022

Gracilienta yingjiangensis Zhang, Meng et Beaver, 2022: 531 [HUYC]

Distribution: P: CH (SW)

Genus *Infrantenna* Liu et Sittichaya, 2022

Infrantenna Liu et Sittichaya, 2022: 46

Type species: *Infrantenna fissilis* Liu et Sittichaya, 2022

Infantenna fissilis Liu et Sittichaya, 2022

Infantenna fissilis Liu et Sittichaya, 2022: 47 [BMNH]

Distribution: I: TH

Genus *Mesoxylion* Vrydagh, 1955

Mesoxylion Vrydagh, 1955a: 19

Type species: *Apate collaris* Erichson, 1842

Mesoxylion collaris (Erichson, 1842)

Apate collaris Erichson, 1842: 148 [ZMUB]

= *Apate bicolor* Lea, 1894: 319 [SAMA]

= *Apate excavata* Lea, 1894: 318 [SAMA]

Distribution: A: AS (ASE ASS AST) NZ

Mesoxylion cylindricus (MacLeay, 1873)

Bostrychus cylindricus MacLeay, 1873: 277 [MAMU]

Distribution: A: AS (ASE ASN ASQ ASS AST) NZ

Mesoxylion perarmatus (Lesne, 1901)

Xylion perarmatus Lesne, 1901a: 562 [MNHN]

Distribution: A: AS (ASE)

Genus *Octodesmus* Lesne, 1901

Octodesmus Lesne, 1901a: 616

Type species: *Octodesmus episternalis* Lesne, 1901

Octodesmus episternalis Lesne, 1901

Octodesmus episternalis Lesne, 1901: 618

Distribution: N: BM ID TH; **Nt_i:** US (USG); **P:** CH (SW)

Octodesmus parvulus (Lesne, 1897)

Xylopertha parvula Lesne, 1897b: 19 [MNHN]

Distribution: I: BM ID (IDC) TH; **P:** GB_i CH (SW) ID (UP)

Genus *Octomeristes* Liu et Beaver in Liu, Beaver et Sanguansub, 2016

Octomeristes Liu et Beaver in Liu, Beaver et Sanguansub, 2016: 3

Type species: *Octomeristes pusillus* Liu et Beaver in Liu, Beaver et Sanguansub, 2016

Octomeristes minutissimus (Lesne, 1932)

Octodesmus minutissimus Lesne, 1932b: 662 [MNHN IFRI]

Distribution: I: ID (IDC, IDS IDW); **P:** ID (UP)

Octomeristes pusillus Liu et Beaver in Liu, Beaver et Sanguansub, 2016

Octomeristes pusillus Liu et Beaver in Liu, Beaver et Sanguansub, 2016: 6 [BMNH]

Distribution: I: TH; **P:** CH (SW)

Genus *Paraxylion* Lesne, 1941

Paraxylion Lesne, 1941b: 150

Type species: *Xylion bifer* Lesne, 1932

Paraxylion bifer (Lesne, 1932)

Xylion bifer Lesne, 1932b: 659 [MNHN IFRI NMPC]

= *Octodesmus kamoli* Chûjô, 1964: 209 [KUFA]

= *Xylion bifer* race *javanicus* Lesne, 1932b: 662 [MNHN IPPB ISNB] [IN]

Distribution: I: BM IA (IAJ) ID (IDS IDW) LO MY (MYC) TH VT; **P:** CH (HKG SW) ID (HP UP)

Genus *Plesioxylion* Liu et Beaver, 2023

Plesioxylion Liu et Beaver, 2023: 22

Type species: *Amintinus gambianus* Borowski, 2018

Plesioxylion gambianus (Borowski, 2018)

Amintinus gambianus Borowski, 2018: 7 [DFPC]

Distribution: E: GA ML SG

Genus *Plioxylion* Vrydagh, 1955

Plioxylion Vrydagh, 1955a: 19

Type species: *Xylopertha plurispinis* Lesne, 1895

Plioxylion plurispinus (Lesne, 1895)

Xylopertha plurispinis Lesne, 1895a: 177 [MNHN]

Distribution: E: MW MZ NB SF TZ UG

Genus *Psicula* Lesne, 1941

Psicula Lesne, 1941b: 146

Type species: *Psicula heterogama* Lesne, 1941

Psicula heterogama Lesne, 1941

Psicula heterogama Lesne, 1941b: 147 [MNHN IFRI]

Distribution: I: ID (IDB); **P:** ID (SD UP)

Genus *Scobicia* Lesne, 1901

Scobicia Lesne, 1901a: 584

Type species: *Apate chevrieri* A. Villa et G. B. Villa, 1835

Scobicia arizonica Lesne, 1907

Scobicia arizonica Lesne, 1907a: 244 [MNHN]

Distribution: **N:** US (USO)

Scobicia barbata (Wollaston, 1860)

Ennaedesmus barbatus Wollaston, 1860b: 539 [OXUM]

Distribution: **N_i:** US; **P:** AZ MR

Scobicia barbifrons (Wollaston, 1864)

Xylopertha barbifrons Wollaston, 1864: 252 [BMNH]

Distribution: **P:** CI

Scobicia bidentata (Horn, 1878)

Sinoxylon bidentatum Horn, 1878: 542 [PANS]

Distribution: **N:** CN (CNL) US (USD USE USL UST)

Scobicia chevrieri (A. Villa et G. B. Villa, 1835)

Apate chevrieri A. Villa et G. B. Villa, 1835: 49 [MSNM?]

= *Xylopertha barbifrons* Walker, 1871: 14 [BMNH] [**HN**]

= *Xylopertha foveicollis* Allard, 1869: 468 [MNHN]

Distribution: **N_i:** CN US (USE UST); **P:** AB AG AL AU BH BU CR CY EG FR GG GR HU IN IS IT LB LE MA MO PT RO RU (CT ST) SK SL SP SY SZ TR TU UK

Scobicia declivis (LeConte, 1859)

Sinoxylon declive LeConte, 1859: 48 [MCZC]

Distribution: **A_i:** HI; **N:** CN (CNN) US (USC USN)

Scobicia ficicola (Wollaston, 1865)

Xylopertha ficicola Wollaston, 1865: 231 [BMNH]

Distribution: **P:** CI

Scobicia lesnei Fisher, 1950

Scobicia lesnei Fisher, 1950: 106 [USNM]

Distribution: **N:** US (USO)

Scobicia monticola Fisher, 1950

Scobicia monticola Fisher, 1950: 108 [USNM]

Distribution: **N:** US (USO)

Scobicia pustulata (Fabricius, 1801)

Apate pustulata Fabricius, 1801: 381 [ZMUC]

= *Apate humeralis* Dejean, 1821: 101 [MNHN?] [**NN**]

= *Apate humeralis* Dejean, 1835: 309 [MNHN?] [**NN HN**]

= *Apate humeralis* Lucas, 1843b: 25 [MNHN] [**HN**]

Distribution: **I$_i$**: ID; **P:** AB AG AU BH CR CY EG FR GE GG GR HU IN IT LB MO PT RO RU (ST) SP SY SZ TR TU UK

Scobicia suturalis (Horn, 1878)

Sinoxylon suturale Horn, 1878: 542 [CASC]

Distribution: **N:** US (USC)

Genus *Sifidius* Borowski et Węgrzynowicz, 2007

Sifidius Borowski et Węgrzynowicz, 2007a: 134

Type species: *Sifidius confusus* Lesne, 1939

= *Sifidius* Lesne, 1939b: 112 [**NN**]

Type species: not designated

Sifidius confusus Lesne, 1939

Sifidius confusus Lesne, 1939b: 113 [BMNH]

Distribution: **Nt:** GT

Sifidius dampfi Lesne, 1939

Sifidius dampfi Lesne, 1939b: 114 [DEIC? MNHN?]

Distribution: **Nt:** MX (MXT MXW)

Genus *Tetrapriocera* Horn, 1878

Tetrapriocera Horn, 1878: 541

Type species: *Tetrapriocera schwarzi* Horn, 1878

Tetrapriocera caprina Lesne, 1931

Tetrapriocera caprina Lesne, 1931a: 104 [MNHN MACN]

Distribution: **Nt:** AY

Tetrapriocera defracta Lesne, 1901

Tetrapriocera defracta Lesne, 1901a: 487 [MNHN]

= *Tetrapriocera laevifrons* Lesne, 1901a: 489 [MNHN]

Distribution: **Nt:** AY (AYN) BZ (BZC BEZ BZM BZS) PG VE

Tetrapriocera longicornis (Olivier, 1790)

Bostrichus longicornis Olivier, 1790b: 15 [MNHN]

= *Tetrapriocera schwarzi* Horn, 1878: 545 [PANS]

Distribution: **A_i:** FPS; **N:** US (USC USE USO); **Nt:** BI BL BR BZ CB CC CU DO DR EC ES GI GL GN GT HA HO MNT MX (MXM MXT) NG PE PG PN PR SN STI STT VE VI; **P_i:** GE

Tetrapriocera oceanina Lesne, 1901

Tetrapriocera oceanina Lesne, 1901a: 486 [MNHN]

Distribution: **A:** FPS (FPM)

Genus *Xylion* Lesne, 1901

Xylion Lesne, 1901a: 542

Type species: *Xylion securifer* Lesne, 1901

Xylion adustus (Fåhraeus, 1871)

Xylopertha adusta Fåhraeus, 1871: 667 [NHRS]

= *Apate excisa* Dejean, 1835: 309 [MNHN?] [NN]

= *Xylopertha pierronii* Fairmaire, 1880a: 246 [MNHN]

Distribution: **E:** AA CG CM KY MB MZ RW SF SO SU TZ UG ZA ZB ZI; **P_i:** GB GE PT

Xylion falcifer Lesne, 1901

Xylion falcifer Lesne, 1901a: 549 [ISNB MNHN SAMC]

Distribution: **E:** AA CF CG EQ KY LH MB MW MZ SF SWA TZ ZA

Xylion inflaticauda Lesne, 1901

Xylion inflaticauda Lesne, 1901a: 553 [ISNB]

Distribution: **E:** AA CG CK EQ GO ZA

Xylion laceratus Lesne, 1901

Xylion laceratus Lesne, 1901a: 548 [ISNB RMNH]

Distribution: **E:** AA MZ NX TZ ZA

Xylion medius Lesne, 1923

Xylion medius Lesne, 1923a: 59 [MNHN MRAC]

Distribution: **E:** AA CG GO RW ZA

Xylion securifer Lesne, 1901

Xylion securifer Lesne, 1901a: 554 [MNHN]

Distribution: **E:** BN CK EQ GH GX IV LI NX SR; **N_i:** US (UST)

Xylion senegambianus Lesne, 1923

Xylion senegambianus Lesne, 1923a: 59 [MNHN]

Distribution: E: AA CK CV GX NX SG

Genus *Xylionopsis* Lesne, 1937

Xylionopsis Lesne, 1937a: 387

Type species: *Xylionopsis urkerewana* Lesne, 1937

Xylionopsis browni Vrydagh, 1959

Xylionopsis browni Vrydagh, 1959c: 10 [ISNB]

Distribution: E: UG

Xylionopsis matruelis Damoiseau, 1968

Xylionopsis matruelis Damoiseau, 1968a: 307 [MRAC ISNB]

Distribution: E: CG ZA

Xylionopsis urkerewana Lesne, 1937

Xylionopsis urkerewana Lesne, 1937a: 387 [ISNB]

= *Xylionopsis ukerewana* Basilewsky, 1952: 135 [LC]

Distribution: E: BD KY RW TZ UG ZA

Genus *Xylionulus* Lesne, 1901

Xylionulus Lesne, 1901a: 516

Type species: *Xylopertha transvena* Lesne, 1900b

Xylionulus epigrus Lesne, 1906

Xylionulus epigrus Lesne, 1906a: 409 [MNHN BMNH]

Distribution: E: MW MZ TZ

Xylionulus maynei Basilewsky, 1954

Xylionulus maynei Basilewsky, 1954: 77 [MRAC]

Distribution: E: RW ZA

Xylionulus pusillus (Fåhraeus, 1871)

Xylopertha pusilla Fåhraeus, 1871: 667 [NHRS]

= *Apate ustulata* Dejean, 1835: 309 [MNHN?] [NN]

Distribution: E: KY NB MB MZ SF

Xylionulus transvena (Lesne, 1900)

Xylopertha transvena Lesne, 1900b: 426 [MNHN SAMC ISNB]

Distribution: E: AA BD CG EH EQ KY MZ NB NX SF STI ZA ZB; **Nt_i:** BZ

Genus *Xylobiops* Casey, 1898

Xylobiops Casey, 1898: 66

Type species: *Apate basilaris* Say, 321

Xylobiops basilaris (Say, 1824)

Apate basilaris Say, 1824: 321 [USNM]

= *Apate basalis* Dejean, 1835: 309 [MNHN?] [**NN**]

= *Apate humeralis* Knoch in Say, 1824: 321 [**NN HN**]

= *Apate humeralis* Melsheimer, 1806: 7 [**NN**]

Distribution: **A**$_i$: NZ; **N:** CN (CNL) US (USD USE USL USO UST); **Nt:** PG; **P**$_i$: GB GE

Xylobiops concisus Lesne, 1901

Xylobiops concisus Lesne, 1901a: 509 [MNHN]

Distribution: **Nt:** CB VE

†*Xylobiops lacustre* Wickham, 1912

Xylobiops lacustre Wickham, 1912: 21 [USNM]

Distribution: **N:** US (USU – Miocene, Florissant)

Xylobiops parilis Lesne, 1901

Xylobiops parilis Lesne, 1901a: 510 [MNHN ZMUB]

Distribution: **N:** MX (MXC MXM MXS) US (USO); **Nt:** CC DR JC

Xylobiops sextuberculatus (LeConte, 1858)

Sinoxylon sextuberculatus LeConte, 1858: 73 [MCZC]

Distribution: **N:** MX (MXC) US (USO USC); **Nt:** GN GT MX (MXM)

Xylobiops texanus (Horn, 1878)

Sinoxylon texanus Horn, 1878: 542 [PANS]

Distribution: **N:** MX (MXC) US (USO); **Nt:** JC

Genus *Xyloblaptus* Lesne, 1901

Xyloblaptus Lesne, 1901a: 476

Type species: *Sinoxylon quadrispinosus* LeConte, 1866

Xyloblaptus mexicanus Lesne, 1939

Xyloblaptus mexicanus Lesne, 1939b: 119 [DEIC? MNHN?]

Distribution: **Nt:** MX (MXT)

Xyloblaptus prosopidis Fisher, 1950

Xyloblaptus prosopidis Fisher, 1950: 131 [USNM]

Distribution: **N:** US (USC)

Xyloblaptus quadrispinosus (LeConte, 1866)
Sinoxylon quadrispinosus LeConte, 1866: 100 [MCZC]
Distribution: **N:** MX (MXC) US (USC USO); **Nt:** GT

Genus *Xylobosca* Lesne, 1901
Xylobosca Lesne, 1901a: 564
Type species: *Bostrychus bispinosa* MacLeay, 1873

Xylobosca bispinosa (MacLeay, 1873)
Bostrychus bispinosa MacLeay, 1873: 276 [MAMU]
= *Rhizopertha elongatula* MacLeay, 1873: 276 [MAMU]
Distribution: **A:** AS (ASE ASN ASQ ASS ASW AST)

Xylobosca canina (Blackburn, 1893)
Xylopertha canina Blackburn, 1893: 131 [BMNH]
= *Xylobosca leai* Lesne, 1901a: 570 [MNHN]
Distribution: **A:** AS (ASE ASQ ASS AST)

Xylobosca capitosa Lesne, 1906
Xylobosca capitosa Lesne, 1906a: 417 [BMNH]
Distribution: **A:** AS (ASW)

Xylobosca cuspidata Lesne, 1906
Xylobosca cuspidata Lesne, 1906a: 416 [ZMUH MNHN]
Distribution: **A:** AS (ASE ASQ ASS ASW)

Xylobosca decisa Lesne, 1906
Xylobosca decisa Lesne, 1906d: 190 [ZMUH MNHN]
Distribution: **A:** AS (ASE ASS)

Xylobosca gemina Lesne, 1901
Xylobosca gemina Lesne, 1901a: 569 [MNHN]
Distribution: **A:** AS (ASE ASQ AST ASW)

Xylobosca mystica (Blackburn, 1890)
Xylopertha mystica Blackburn, 1890: 1264 [BMNH]
Distribution: **A:** AS (ASS)

Xylobosca neboissi Vrydagh, 1957
Xylobosca neboissi Vrydagh, 1957: 13 [MMMA ISNB]
Distribution: **A:** AS (ASE)

Xylobosca spinifrons Lesne, 1906

Xylobosca spinifrons Lesne, 1906a: 415 [ZMUH MNHN]

= *Xylobosca geometrica* Lesne, 1906d: 192 [ZMUH]

Distribution: A: AS (ASE ASN ASQ)

Xylobosca vicaria Lesne, 1906

Xylobosca vicaria Lesne, 1906a: 419 [ZMUH MMMA]

Distribution: A: AS (ASE ASQ)

Xylobosca vidua (Blackburn, 1890)

Xylopertha vidua Blackburn, 1890: 1265 [BMNH]

Distribution: A: AS (ASE ASQ ASS)

Genus *Xylocis* Lesne, 1901

Xylocis Lesne, 1901a: 519

Type species: *Xylocis tortilicornis* Lesne, 1901

Xylocis tortilicornis Lesne, 1901

Xylocis tortilicornis Lesne, 1901a: 520 [MNHN ZMUB]

Distribution: I: CX ID (IDH IDS IDW) LO TH; **P:** CH (HKG SW TAI) ID (UP) NP

Genus *Xylodectes* Lesne, 1901

Xylodectes Lesne, 1901a: 536

Type species: *Xylopertha ornata* Lesne, 1897

Xylodectes ornatus (Lesne, 1897)

Xylopertha ornata Lesne, 1897b: 19 [MNHN ISNB]

Distribution: I: BM IA (IAB IAS) ID (IDB IDC IDS IDW) LO PH SX TH VT; **P:** CH (HAI SE SW TAI) ID (SD UP)

Xylodectes venustus Lesne, 1901

Xylodectes venustus Lesne, 1901a: 540 [MNHN]

Distribution: A: AS (ASQ); **I:** VT

Genus *Xylodeleis* Lesne, 1901

Xylodeiles Lesne, 1901a: 524

Type species: *Apate obsipa* Germar, 1848

Xylodeleis obsipa (Germar, 1848)

Apate obsipa Germar, 1848: 222 [ZMUB]

= *Apate retusa* Lea, 1894: 320 [SAMA]

= *Apate serrata* Lea, 1894: 317 [SAMA]
= *Apate subcostata* Lea, 1894: 317 [SAMA]
= *Sinoxylon rufescens* Murray, 1867: 94 [BMNH]
Distribution: A: AS (ASE ASQ ASS AST) NZ; **I_i:** CX; **P_i:** CH (TAI-Lanhsu I.)

Genus *Xylodrypta* Lesne, 1901

Xylodrypta Lesne, 1901a: 523
Type species: *Xylodrypta bostrychoides* Lesne, 1901

Xylodrypta bostrychoides Lesne, 1901

Xylodrypta bostrychoides Lesne, 1901a: 523 [MNHN]
Distribution: I: ID (IDE) TH

Xylodrypta guochuanii Zhang, Meng et Beaver, 2022

Xylodrypta guochuanii Zhang, Meng et Beaver, 2022: 535 [IZAS]
Distribution: P: CH (SW)

Xylodrypta lanna Liu et Beaver in Liu, Beaver et Sanguanshub, 2021

Xylodrypta lanna Liu et Beaver in Liu, Beaver et Sanguanshub, 2021: 236 [BMNH]
Distribution: I: LO TH

Xylodrypta zhenghei Zhang, Meng et Beaver, 2022

Xylodrypta zhenghei Zhang, Meng et Beaver, 222: 537 [IZAS]
Distribution: P: CH (SW)

Genus *Xylogenes* Lesne, 1901

Xylogenes Lesne, 1901a: 614
Type species: *Rhizoperta dilatata* Reitter, 1889

Xylogenes dilatatus (Reitter, 1889)

Rhizoperta dilatata Reitter, 1889: 112 [NHMW]
Distribution: P: AB IN IQ JO KZ KY SY TD TM UZ

Xylogenes granulicauda Lesne, 1941

Xylogenes granulicauda Lesne, 1941a: 132 [BMNH]
Distribution: A: AS (ASQ ASW)

Xylogenes hirticollis (Blackburn, 1897)

Xylopertha hirticollis Blackburn, 1897: 92 [NHRS]
= *Xylogenes mjoebergianus* Lesne, 1932c: 6 [NHRS]
Distribution: A: AS (ASE ASN ASS ASW)

Xylogenes mesopotamicus Lesne, 1937

Xylogenes mesopotamicus Lesne, 1937b: 195 [MNHN DEIC]

Distribution: P: AR IN IQ OM SY

Xylogenes semenovi Lesne, 1904

Xylogenes semenovi Lesne, 1904: 158 [ZMAS]

Distribution: P: CH (NW) TM

Xylogenes sindicola Lesne, 1936

Xylogenes sindicola Lesne, 1936a: 136 [MNHN IFRI]

Distribution: P: PA

Genus *Xylomeira* Lesne, 1901

Xylomeira Lesne, 1901a: 502

Type species: *Apate torquata* Fabricius, 1801

Xylomeira tridens (Fabricius, 1792)

Apate tridens Fabricius, 1792: 362 [ZMUC]

= *Apate planifrons* Lesne, 1901a: 503 [**NN**]

= *Apate torquata* Fabricius, 1801: 382 [ZMUC]

= *Sinoxylon floridanum* Horn, 1885: 155 [PANS]

= *Xyloprista fisheri* Rai, 1978: 119 [IFRI]

Distribution: $\mathbf{I_i}$: ID (IDP); **N:** US (USE USO); **Nt:** AT BI CU DO DR GL GN HA JC LW MNT MQ MX (MXM MXT MXW) PR SN SQ STI STT VI

Genus *Xylopertha* Guérin-Méneville, 1845

Xylopertha Guérin-Méneville, 1845: 17

Type species: *Apate sinuatus* Fabricius, 1792

= *Paraxylogenes* Damoiseau, 1968b: 2

Type species: *Paraxylogenes pistaciae* Damoiseau, 1968b

= *Xylonites* Lesne, 1901a: 575

Type species: *Bostrichus retusus* Olivier, 1790

Xylopertha elegans Liu et Beaver, 2017

Xylopertha elegans Liu et Beaver, 2017: 8 [ZMUB]

Distribution: P: TR

Xylopertha praeusta (Germar, 1817)

Apate praeusta Germar, 1817: 226 [ZMUB]

= *Apate appendiculata* Dejean, 1821: 101 [MNHN?] [**NN**]

= *Apate appendiculata* Dejean, 1835: 309 [MNHN?] [**NN HN**]

= *Apate appendiculata* Dejean in Sturm, 1826: 94 [MNHN?] [**NN HN**]

= *Xylopertha appendiculata* Lucas, 1849: 466 [MNHN] [**HN**]

= *Xylopertha dunensis* Rai et Chatterjee, 1964: 122 [IFRI]

= *Xylopertha humeralis* Gemminger et Harold, 1869: 1790 [**NN**]

Distributon: **I_i**: ID; **P:** AG CR GR FR ID (UP) IS IT MO PT SP TU

Xylopertha reflexicauda (Lesne, 1937)

Xylonites reflexicauda Lesne, 1937b: 199 [MNHN IFRI ISNB]

= *Paraxylogenes pistaciae* Damoiseau, 1968b: 4 [BMNH ISNB]

Distribution: **A_i**: AS; **I:** ID; **N_i**: US; **P:** GR (Crete) IN IQ IS PA TR

Xylopertha retusa (Olivier, 1790)

Bostrichus retusus Olivier, 1790a: 110 [MNHN]

= *Apate aterrima* Faldermann, 1837: 250 [ZMUM?]

= *Apate capillata* Dahl in Dejean, 1835: 309 [MNHN?] [**NN HN**]

= *Apate capillata* Dahl in Sturm, 1826: 94 [unknown] [**NN**]

= *Apate capillata* Dahl in Sturm, 1843: 231 [unknown] [**NN HN**]

= *Apate sinuata* Fabricius, 1792: 362 [MNHN]

= *Apate sinuata* Fabricius in Dejean, 1835: 309 [MNHN?] [**NN HN**]

= *Apate sinuata* Fabricius in Sturm, 1843: 231 [MNHN?] [**NN HN**]

= *Apate sinuata* Megerle in Dejean, 1821: 101 [MNHN] [**NN**]

= *Apate sinuata* Sturm, 1826: 95 [MNHN?] [**NN HN**]

= *Bostrichus retusa* Dejean, 1821: 101 [MNHN] [**NN HN**]

= *Bostrichus retusus* Dejean, 1835: 307 [MNHN] [**NN HN**]

Distribution: **P:** AB AG AL AU BE BH BU CR CY CZ FR GE GG GR HU IN IS IT LB LU MC MD MO NL PL PT RO RU (ES ST WS) SK SL SP SZ TR TU UK YU

Genus *Xyloperthella* Fisher, 1950

Xyloperthella Fisher, 1950: 116

Type species: *Bostrichus piceus* Olivier, 1790

Xyloperthella crinitarsis (Imhoff, 1843)

Apate crinitarsis Imhoff, 1843: 177 [NHMB]

= *Sinoxylon pubescens* Murray, 1867: 93 [BMNH]

Distribution: **E:** AA BD CG CK EQ GH GO GS GX IV KY LI MT NX RW SG SR STI TG TZ UG ZA; **N_i**: US; **P_i**: FR GB GE CH (TAI) NL SP

Xyloperthella picea (Olivier, 1790)

Bostrichus piceus Olivier, 1790a: 110 [MNHN]

= *Apate cultrata* J. Thomson, 1858: 83 [MNHN]

= *Apate femorata* Klug, 1835: 203 [ZMUB]
= *Apate frenata* Klug in Dejean, 1835: 309 [MNHN?] [NN]
= *Apate truncata* Dejean, 1835: 309 [MNHN?] [NN]
= *Sinoxylon fumatum* Murray, 1867: 94 [BMNH]
= *Xylopertha heydeni* Schilsky, 1899: 92 [ZMUB]
= *Xylopertha pica* Schedl, 1964: 285 [LC]
= *Xylopertha picea plumbeipennis* Lesne, 1924: 222 [MNHN MCSN MRAC]
= *Xyloperthella guineensis* Roberts, 1967: 87 [BMNH]

Distribution: E: AA BF CG CK CM CV EH EQ ER GA GH GO GS GX IV KY MB ML MT MU MZ NX RI RW SF SG SSU SO SR STI SU SYC TG TZ UG ZA ZB; N_i: US; Nt_i: AY BZ (BZC) CB FG JC PE PG; P_i: AG EG FR GB GE IS IT LB MA MO NL OM PT SA SP SY TR TU YE (YEC)

Xyloperthella scutula (Lesne, 1901)

Xylopertha scutula Lesne, 1901a: 535 [MNHN MNMS]

Distribution: E: BF CD EH GA NI SG SR SU ZA

Genus *Xylophorus* Lesne, 1906

Xylophorus Lesne, 1906a: 419

Type species: *Xylophorus abnormis* Lesne, 1906

Xylophorus abnormis Lesne, 1906

Xylophorus abnormis Lesne, 1906a: 420 [BMNH]

Distribution: I: CX

Xylophorus ceylonicus Lesne, 1941

Xylophorus ceylonicus Lesne, 1941b: 143 [IFRI]

Distribution: I: CX

Genus *Xyloprista* Lesne, 1901

Xyloprista Lesne, 1901a: 497

Type species: *Xylopertha hexacantha* Fairmaire, 1892

Xyloprista arcellata (Lesne, 1901)

Xylomeira arcellata Lesne, 1901a: 499 [MNHN]

Distribution: Nt: BZ (BZA BZM) PE PG VE

Xyloprista hexacantha (Fairmaire, 1892)

Xylopertha hexacantha Fairmaire, 1892b: 245 [MNHN]

Distribution: N_i: US (USE); Nt: AY (AYN AYS) BV BZ CB PG UR

Xyloprista longiscapus Liu et Beaver, 2013

Xyloprista longiscapus Liu et Beaver, 2013a: 96 [ZSMC]

Distribution: **Nt:** JC

Xyloprista praemorsa (Erichson, 1847)

Xylopertha praemorsa Erichson, 1847: 309 [ZMUB]

= *Apate dentata* Dejean, 1835: 309 [MNHN?] [**NN**]

Distribution: **Nt:** BV BZ (BZS) CB PE PG VE

Genus *Xylopsocus* Lesne, 1901

Xylopsocus Lesne, 1901a: 627

Type species: *Apate capucinus* Fabricius, 1781

Xylopsocus acutespinosus Lesne, 1906

Xylopsocus acutespinosus Lesne, 1906a: 424 [MNHN BMNH]

Distribution: **I:** BM LO TH; **P:** CH (NO SW) ID (SD) NP

Xylopsocus bicuspis Lesne, 1901

Xylopsocus bicuspis Lesne, 1901a: 634 [MNHN]

Distribution: **P:** CH (SW TAI) JA

Xylopsocus burnsi Vrydagh, 1958

Xylopsocus burnsi Vrydagh, 1958a: 60 [MMMA]

Distribution: **A:** AS (ASE)

Xylopsocus capucinus (Fabricius, 1781)

Apate capucina Fabricius, 1781: 62 [BMNH]

= *Apate capucina* Sturm, 1826: 101 [MNHN?] [**NN HN**]

= *Apate fuliginosa* Dejean, 1835: 309 [MNHN?] [**NN**]

= *Apate infuscata* Dejean, 1835: 309 [MNHN?] [**NN**]

= *Apate marginata* Fabricius, 1801: 382 [ZMUC]

= *Apate marginata* Sturm, 1826:94 [MNHN?] [**NN HN**]

= *Bostrichus eremita* Olivier, 1790a: 110 [BMNH]

= *Ennaedesmus nicobaricus* Redtenbacher, 1868: 114 [MNMW?]

Distribution: **A:** AS IA (IAN SH) MI NA NH PI PW SH TRC; **E_i:** CM CV GX KY LI MB MI MU MZ NX RI SF SG SR SYC TZ; **I:** BM CA CX IA (IAB IAC IAJ IAM IAS) ID LO MY (MYC MYS) PH TH VT; **N_i:** US (USE); **Nt_i:** BZ DO DR FG GU HO SM TT VE; **P:** CH (CE HAI SE SW TAI) JA ID (SD) NP

Xylopsocus castanopterus (Fairmaire, 1850)

Apate castanoptera Fairmaire, 1850: 50 [MNHN?]

= *Apate affinis* Brancsik, 1893: 235 [FMNH]

= *Xylopsocus distinctus* Rai, 1967a: 140 [IFRI]

Distribution: A: AS FJ FPS HI IA (SH) NA PW SJ SS TA; **E:** CM KY MB MU RI SF TZ; **I:** AI BG IA (IAB IAN) MY (MYS) PH SX VT; **N_i:** US (UST); **P:** AU CH (TAI) ID (UP) JA (including Bonin and Volcano Is.)

Xylopsocus ebeninocollis Lesne, 1901

Xylopsocus ebeninocollis Lesne, 1901a: 631 [MCSN]

Distribution: A: PW

Xylopsocus edentatus (Montrouzier, 1861)

Apate edentatus Montrouzier, 1861: 267 [MNHN?]

Distribution: A: NA

Xylopsocus ensifer Lesne, 1906

Xylopsocus ensifer Lesne, 1906a: 425 [BMNH]

Distribution: I: BM MY (MYC) TH

Xylopsocus galloisi Lesne, 1937

Xylopsocus galloisi Lesne, 1937b: 198 [MNHN]

Distribution: P: CH (NO SW) JA SC

Xylopsocus gibbicollis (MacLeay, 1873)

Rhizopertha gibbicollis MacLeay, 1873: 276 [MAMU]

Distribution: A: AS (ASE ASQ ASS AST ASW) NZ PW

Xylopsocus indianus Vrydagh, 1959

Xylopsocus indianus Vrydagh, 1959c: 13 [ISNB]

Distribution: I: ID (IDS)

Xylopsocus intermedius Damoiseau in Damoiseau et Coulon, 1993

Xylopsocus intermedius Damoiseau in Damoiseau et Coulon, 1993: 53 [HNHM]

Distribution: I: VT; **P:** CH (NW SW)

Xylopsocus philippinensis Vrydagh, 1955

Xylopsocus philippinensis Vrydagh, 1955c: 8 [ISNB]

Distribution: I: PH; **P:** CH (CE)

Xylopsocus radula Lesne, 1901

Xylopsocus radula Lesne, 1901a: 634 [MNHN]

Distribution: I: BM IA (IAS) ID MY (MYC) TH VT; **P:** CH (SW)

Xylopsocus ritsemai Lesne, 1906

Xylopsocus ritsemai Lesne, 1906a: 422 [RMNH]

Distribution: I: IA (IAJ)

Xylopsocus rubidus Lesne, 1901

Xylopsocus rubidus Lesne, 1901a: 629 [MNHN]

Distribution: A: AS (ASE ASS ASW)

Xylopsocus sellatus (Fåhraeus, 1871)

Xylopertha sellatus Fåhraeus, 1871: 667 [NHRS]

Distribution: E: CG KY MB RI RW SF TZ UG ZA; **P:** CH (TAI)

Genus *Xylothrips* Lesne, 1901

Xylothrips Lesne, 1901a: 620

Type species: *Apate flavipes* Illiger, 1801

Subgenus *Xylothrips* Lesne, 1901

Xylothrips Lesne, 1901a: 620

Type species: *Apate flavipes* Illiger, 1801

Xylothrips* (*Xylothrips*) *flavipes (Illiger, 1801)

Apate flavipes Illiger, 1801: 171 [ZMUB]

= *Apate dominicana* Fabricius, 1801: 380 [ZMUC]

= *Apate macrocera* Latrreille in Dejean, 1835: 309 [MNHN?] [**NN**]

= *Apate sinuata* Stephens, 1830: 351 [BMNH]

= *Bostrichus iracundus* Snellen van Vollenhoven in Snellen van Vollenhoven et Sélys Longchamps, 1869: 10 [RMNH]

= *Bostrichus mutilatus* Walker, 1858b: 286 [BMNH]

= *Bostrichus mutilatus* Walker, 1859: 260 [BMNH] [**HN**]

Distribution: A$_i$: AS (ASE) FPS IA (IAN SH) MI PI; **E:** CM MB MAY MU NX RI SF SYC TZ; **I:** AI BM CX IA (IAB IAC IAJ IAS) ID (IDB) LO MY (MYC) PH TH VT; **N$_i$:** US (USC UST); **Nt$_i$:** MQ; **P$_i$:** AG BT CH (HAI NO SE SW TAI) FR GB GR ID (SD UP) IS IT JA NP SA YE (YES)

Xylothrips* (*Xylothrips*) *religiosus (Boisduval, 1835)

Apate religiosa Boisduval, 1835: 460 [BMNH?]

= *Apate destructor* Montrouzier, 1857: 55 [MNHN?]

= *Apate geoffroyi* Montrouzier, 1861: 266 [MNHN?]

= *Apate lifuana* Montrouzier, 1861: 267 [MNHN?]

= *Apate religiosa* Dejean, 1821: 101 [MNHN?] [**NN**]

= *Apate religiosa* Dejean, 1835: 309 [MNHN?] [**NN HN**]

= *Apate religiosae* Fairmaire, 1850: 50 [**LC**]

Distribution: A: AS (ASN ASE ASQ) CIS CO FJ FPS (FPM) HI IA (SH) NA NH PI PW SJ SS TA; **E_i:** MU RI; **I:** IA (IAC IAN) MY; **N_i:** US (UST); **P_i:** AU CH (NO SW TAI)

Sugenus *Calophagus* Lesne, 1902

Calophagus Lesne, 1902a: 108

Type species: *Calophagus pekinensis* Lesne, 1902a

Xylothrips (*Calophagus*) *colombiana* Zhang, Meng et Beaver, 2022

Calophagus colombiana Zhang, Meng et Beaver, 2022: 529 [IZAS]

Distribution: P: CH (SW)

Xylothrips (*Calophagus*) *pekinensis* Lesne, 1902

Calophagus pekinensis Lesne, 1902a: 109 [MNHN]

= *Calophagus pekinensis* Liu, 2021a: 141 [**HN**]

= *Calophagus perinensis* Hua, 2002: 150 [**LC**]

= *Xylothrips cathaicus* Reichardt, 1966: 82 [MCZC MZSP]

Distribution: P: CH (CE NO) JA SC

Genus *Xylotillus* Lesne, 1901

Xylotillus Lesne, 1901a: 540

Type species: *Apate lindi* (Blackburn, 1890)

Xylotillus lindi (Blackburn, 1890)

Apate lindi Blackburn, 1890: 1263 [BMNH]

Distribution: A: AS (ASE ASN ASS)

Subfamily Dinoderinae C. G. Thomson, 1863

Dinoderina C. G. Thomson, 1863: 201

Type genus: *Dinoderus* Stephens, 1830

Genus *Dinoderopsis* Lesne, 1906

Dinoderopsis Lesne, 1906a: 400

Type species: *Dinoderopsis escharipora* Lesne, 1906

Dinoderopsis eschariporia Lesne, 1906

Dinoderopsis eschariporia Lesne, 1906a: 401 [BMNH]

Distribution: P: YE (YES)

Dinoderopsis opimus Lesne, 1938

Dinoderopsis opimus Lesne, 1938b: 174 [BMHN, ISNB]

Distribution: E: SF

Dinoderopsis serriger Lesne, 1923

Dinoderopsis serriger Lesne, 1923a: 56 [MNHN, BMNH]

Distribution: E: AA CD MZ UG ZA ZI; **I:** LO TH; **P:** YE (YEC YES)

Genus *Dinoderus* Stephens, 1830

Dinoderus Stephens, 1830: 352

Type species: *Dinoderus ocellaris* Stephens, 1830

= *Patea* Casey, 1898: 66

Type species: *Dinoderus brevis* Horn, 1878

Subgenus *Dinoderastes* Lesne, 1914

Dinoderastes Lesne, 1914a: 244

Type species: *Dinoderus scabricauda* Lesne, 1914

Dinoderus (*Dinoderastes*) *exilis* Lesne, 1932

Dinoderus exilis Lesne, 1932b: 653 [MNHN]

Distribution: I: IA (IAC) ID (IDB IDE) TH; **P:** BT ID (SD)

Dinoderus (*Dinoderastes*) *hongheensis* Zhang, Meng et Beaver, 2022

Dinoderus (*Dinoderastes*) *hongheensis* Zhang, Meng et Beaver, 2022: 508 [IZAS]

Distribution: P: CH (SW)

Dinoderus (*Dinoderastes*) *japonicus* Lesne, 1895

Dinoderus japonicus Lesne, 1895a: 170 [MNHN]

= *Dinoderus pubicollis* Van Dyke, 1923: 45 [CASC]

= *Dinoderus tsugae* Matsumura, 1915b: 183 [EIHU]

Distribution: **A_i:** AS (ASE); **N_i:** CN US (USC USD USG USO); **P:** AU_i GB_i GE_i CH (CE HKG NO SE SW TAI) FR_i IT_i JA NL_i PL_i RU_i (CT) SC SV_i SZ_i

Dinoderus (*Dinoderastes*) *nanxiheensis* Zhang, Meng et Beaver, 2022

Dinoderus (*Dinoderastes*) *nanxiheensis* Zhang, Meng et Beaver, 2022: 510 [IZAS]

Distribution: P: CH (SW)

Dinoderus (*Dinoderastes*) *scabricauda* Lesne, 1914

Dinoderus scabricauda Lesne, 1914a: 243 [MNHN]

Distribution: I: PH

Dinoderus (*Dinoderastes*) *speculifer* Lesne, 1895

Dinoderus speculifer Lesne, 1895a: 169 [MNHN]

Distribution: I: IA (IAC); **P:** CH (NE SW TAI) JA SC

Subgenus *Dinoderus* Stephens, 1830

Dinoderus Stephens, 1830: 352

Type species: *Dinoderus ocellaris* Stephens, 1830

Dinoderus (*Dinoderus*) *bifoveolatus* (Wollaston, 1858)

Rhyzopertha bifoveolata Wollaston, 1858: 409 [MNHN, ISNB]

= *Dinoderus bifoieolatus* Vrydagh, 1962: 7 [LC] misspelling and error in author of description – Vrydagh (1962) mentioned Horn (1878: 549, 550), as the author of this species

= *Dinoderus perpunctatus* Lesne, 1895a: 170

Distribution: A: IA (IAN) MI NA PW; **E:** AA BN CG CV GA GH GX IV MB MT MU NX SG SR TG TZ ZA; **I:** IA (IAB IAJ IAS) ID MY (MYC MYS) PH TH VT; **N_i:** CN (CNE) US (USO); **Nt_i:** AY (AYN) BZ (BZA) CU FG GL GN GR GU HA HO PE PG STV VE; **P_i:** AU BE CR CH (SE TAI) CZ DE EG FI FR GB GE IS IT JA MR NL NP PA PL SA SK SP SZ YE (YEC)

Dinoderus (*Dinoderus*) *borneanus* Lesne, 1933

Dinoderus borneanus Lesne, 1933a: 257 [BMNH]

Distrubution: I: IA (IAB) MY (MYC)

Dinoderus (*Dinoderus*) *brevis* Horn, 1878

Dinoderus brevis Horn, 1878: 549 [PANS]

Distribution: A: IA (SH); **E_i:** CG MZ TZ ZA; **I:** BG BM IA (IAB IAJ IAN) ID MY (MYC) PH TH VT; **N_i:** US (USL USN USO UST); **Nt_i:** JC PG; **P:** AZ_i FI_i FR_i GB_i GE_i GR_i CH (TAI) NL_i NP PL_i RU_i (NT) SV_i

Dinoderus (*Dinoderus*) *creberrimus* Lesne, 1941

Dinoderus creberrimus Lesne, 1941b: 141 [MNHN, IFRI]

Distribution: I: ID (IDE); **P:** CH (SW)

Dinoderus (*Dinoderus*) *distinctus* Lesne, 1898

Dinoderus distinctus Lesne, 1898c: 325 [MNHN]

Distribution: E_i: CG KY TZ ZA; **I:** PH VT; **Nt_i:** GL; **P:** GE_i CH (TAI)

Dinoderus* (*Dinoderus*) *favosus Lesne, 1911

Dinoderus favosus Lesne, 1911f: 397 [MNHN]

Distrbution: I: AI BM CA ID (IDB IDE) TH VT; **P:** CH (SW)

Dinoderus* (*Dinoderus*) *gabonicus Lesne, 1921

Dinoderus gabonicus Lesne, 1921a: 132 [MNHN]

Distribution: E: GO KY

Dinoderus* (*Dinoderus*) *gardneri Lesne, 1933

Dinoderus gardneri Lesne, 1933a: 258 [IFRI, MNHN, ISNB]

Distribution: I: ID (IDS IDW) TH

Dinoderus* (*Dinoderus*) *glabripennis Lesne, 1911

Dinoderus glabripennis Lesne, 1911a: 398 [BMNH]

Distribution: I: BM

Dinoderus* (*Dinoderus*) *koi Borowski et Węgrzynovicz, 2013

Dinoderus koi Borowski et Węgrzynovicz, 2013: 34 [NTUC]

Distribution: P: CH (TAI)

Dinoderus* (*Dinoderus*) *mangiferae Lesne, 1921

Dinoderus mangiferae Lesne, 1921a: 132

Distribution: I: ID (IDB); **P:** ID (UP)

Dinoderus* (*Dinoderus*) *minutus (Fabricius, 1775)

Apate minutus Fabricius, 1775: 54 [BMNH]

= *Apate pumila* Dejean, 1821: 101 [MNHN?] [**NN**]

= *Apate pumila* Dejean, 1835: 309 [MNHN?] [**NN HN**]

= *Apate umbilicatus* Mannerheim in Dejean, 1835: 309 [MNHN?] [**NN**]

= *Bostrichus vertens* Walker, 1859: 260 [BMNH]

= *Rhizopertha sicula* Baudi di Selve, 1874: 336 [BMNH]

= *Xylopertha bambulae* Dalman in Gemminger et Harold, 1869: 1790 [**NN**]

Distribution: $\mathbf{A_i}$: AS (ASE) CO FPS (FPM) IA (IAN SH) MI NA NZ PI PW SH TA TI; $\mathbf{E_i}$: BD BN CG GA GH GX IV KY MB MU MW MZ NI NX RI RW SF SG SU SWA SYC TZ UG ZA ZB; **I:** BM CHI CX IA (IAC IAJ IAM IAS) ID (IDC IDS) MY (MYC MYS) PH TH VT; $\mathbf{N_i}$: CN (CNL CNM CNN CNP) US (USC USE USM); $\mathbf{Nt_i}$: AY (AYN) BR BZ (BZC) CB CU DO DR EC GI GL GN GT HA HO JC MNT MX (MXT) NG PG PR STV VE VI; **P:** AG_i AU_i BE_i CH (CE HAI NO SE SW TAI) CY_i CZ_i DE_i EG_i FI_i FR_i GB_i GE_i GR_i HU_i IS_i IT_i JA MO_i NL_i NR_i PL_i SC_i SK_i SP_i SV_i SY_i SZ_i

Dinoderus (Dinoderus) nitidus Lesne, 1898

Dinoderus nitidus Lesne, 1898c: 321 [MNHN]

Distribution: A: FPS (FPM) PW

Dinoderus (Dinoderus) oblongopunctatus Lesne, 1923

Dinoderus oblongopunctatus Lesne, 1923a: 55 [MNHN]

Distribution: E: CG GX TZ ZA

Dinoderus (Dinoderus) ocellaris Stephens, 1830

Dinoderus ocellaris Stephens, 1830: 352 [OXUM]

= *Dinoderus australiensis* Lesne, 1897e: 184 [RMNH]

= *Dinoderus pilifrons* Lesne, 1895a: 170 [MNHN]

Distribution: A: AS (ASE ASN) IA (IAN) NZ PW; E: CG GA ZA; I: BM CA CX IA (IAC IAJ IAJ – Lombok I.) IAS) ID (IDC IDE) LO MY (MYC) PH PW TH VT; N_i: US (USE USN USO UST); P_i: BE BH CH (CE SE TAI) FR GB GE NL NP PL SP SV

Dinoderus (Dinoderus) ochraceipennis Lesne, 1906

Dinoderus ochraceipennis Lesne, 1906a: 399 [MNHN BMNH]

Distribution: I: BM ID VT; P: CH (SW)

Dinoderus (Dinoderus) papuanus Lesne, 1899

Dinoderus papuanus Lesne, 1899d: 636 [MNHN]

Distrubution: A: AS (ASE) PW

Dinoderus (Dinoderus) perfoliatus Gorham, 1886

Dinoderus perfoliatus Gorham, 1886: 353 [MNHN BMNH]

Distribution: Nt: PN

Dinoderus (Dinocerus) perplexus Lesne, 1932

Dinoderus perplexus Lesne, 1932b: 651 [IFRI]

Distribution: I: ID (IDS IDW)

Dinoderus (Dinoderus) piceolus Lesne, 1933

Dinoderus piceolus Lesne, 1933a: 259 [BMNH]

Distribution: P: CH (HKG)

Dinoderus (Dinoderus) politulus Lesne, 1941

Dinoderus politulus Lesne, 1941b: 140 [IFRI?]

Distribution: I: ID (IDS)

Dinoderus* (*Dinoderus*) *porcellus Lesne, 1923

Dinoderus porcellus Lesne, 1923a: 55 [MNHN]

Distribution: E: GX IV NX SR

Dinoderus* (*Dinoderus*) *punctatissimus Lesne, 1898

Dinoderus punctatissimus Lesne, 1898c: 329 [MNHN]

Distribution: I: ID (IDS, „Hindoustan")

Dinoderus **Lesne, 1906, subgenus incertae sedis**

†***Dinoderus cuneicollis*** Wickham, 1913

Dinoderus cuneicollis Wickham, 1913: 16 [USNM]

Distribution: N: US (USU – Miocene, Florissant)

Genus *Orientoderus* Borowski et Węgrzynowicz, 2011

Prostephanus (*Orientoderus*) Borowski et Węgrzynowicz, 2011b: 255

Type species: *Prostephanus* (*Orientoderus*) *orientalis* Borowski et Węgrzynowicz, 2011

Orientoderus orientalis Borowski et Węgrzynowicz, 2011

Prostephanus orientalis Borowski et Węgrzynowicz, 2011b: 256 [HNHM]

Distribution: I: LO TH; **P:** CH (SW)

Genus *Prostephanus* Lesne, 1898

Prostephanus Lesne, 1898c: 342

Type species: *Apate punctata* Say, 1826

Prostephanus apax Lesne, 1930

Prostephanus apax Lesne, 1930: 102 [MNHN]

Distribution: N: US (USO); **Nt:** MX (MXE) PN

Prostephanus arizonicus Fisher, 1950

Prostephanus arizonicus Fisher, 1950: 38 [USNM]

Distribution: N: US (USO)

Prostephanus mexicanus Liu et Beaver, 2013

Prostephanus mexicanus Liu et Beaver, 2013b: 258 [CACM]

Distribution: N: US (USE); **Nt:** MX (MXM)

Prostephanus punctatus (Say, 1826)

Apate punctatus Say, 1826: 258 [USNM]

= *Apate punctatus* Melsheimer, 1806: 7 [NN]

Distribution: N: CN (CNL CNM) US (USD USE USG USO UST); **Nt:** GT MX (MXM) NG

Prostephanus sulcicollis (Fairmaire in Fairmaire et Germain, 1861)

Xylopertha sulcicollis Fairmaire in Fairmaire et Germain, 1861: 4 [USNM]

Distributions: Nt: CL (CLN, Juan Fernandes I.)

Prostephanus truncatus (Horn, 1878)

Dinoderus truncatus Horn, 1878: 549 [PANS]

= *Prostephanus trucatus* Hua, 2002: 151 [**LC**]

Distributions: $\mathbf{E_i}$: BD BN BF CG EH GH GT GX KY MW MZ NB NI NX RW SF TG TZ UG ZA ZI; $\mathbf{I_i}$: ID PH TH; **N:** CN MX (MXC) US (USC USO UST); **Nt:** BZ CB CC GT HO MX (MXM MXS MXT MXW) NG PA PE Salvador; $\mathbf{P_i}$: FR GE CH (CE SE SW) IQ IS

Genus *Rhizoperthodes* Lesne, 1936

Rhizoperthodes Lesne, 1936a: 134

Type species: *Rhizoperthodes anguinus* Lesne, 1936

Rhizoperthodes anguinus Lesne, 1936

Rhizoperthodes anguinus Lesne, 1936a: 134 [MNHN BMNH]

Distribution: I: MY

Genus *Rhyzopertha* Stephens, 1830

Rhyzopertha Stephens, 1830: 354

Type species: *Synodendron pusillum* Fabricius, 1798

= *Rhizopertha* Guérin-Méneville, 1845: 17 [**LC**]

Rhyzopertha dominica (Fabricius, 1792)

Synodendron dominicum Fabricius, 1792: 359 [ZMUC]

= *Apate castanea* Ullrich in Dejean, 1835: 309 [MNHN?] [**NN HN**]

= *Apate castanea* Ullrich in Sturm, 1826: 94 [unknown] [**NN**]

= *Apate castanea* Ullrich in Sturm, 1843: 231 [unknown] [**NN HN**]

= *Apate frumentaria* Nördlinger, 1855: 189 [unknown]

= *Apate rufa* Hope, 1845: 17 [BMNH]

= *Bostrichus moderatus* Walker, 1859: 260 [BMNH]

= *Ptinus fissicornis* Marsham, 1802: 82 [BMNH]

= *Ptinus piceus* Marsham, 1802: 88 [BMNH]

= *Rhyzopertha dominica* forma *granulipennis* Lesne in Beeson et Bhatia, 1937: 283 [**IN**]

= *Rhyzopertha hordeum* Matsumura in Hagstrum et Subramanyam, 2009: 160 [**NN**] **(syn. nov.)**

= *Synodendron pusillum* Fabricius, 1798: 156 [ZMUC?]

Distribution: $\mathbf{A_i}$: AS (ASE ASN ASW) CO FJ HI MI NA NH NZ PI PW SJ; $\mathbf{E_i}$: AA BD BF BW CG CV DJ EH EQ GO GX IV KY MB ML MU MW MZ NI NX RI RW SF SG SR SO STI SU SVH SWA TZ UG ZA ZB; **I:** AI BG BM CX IA (IAJ) ID (IDB IDC IAJ IDS) LO MY (MYC) PH SIG SK TH VT; $\mathbf{N_i}$: CN (CNL

CNM CNN CNP) US (USD USE USG USL USN USO); **Nt_i:** AY (AYS) BR BZ CB CL (CLN) CU GL GT HA HO EC JC MQ MX (MXT) NG PE PG PR VE VI; **P_i:** AF AG AR AU AZ BE BT BU BY CI CR CY CZ DE EG FI FR GB GE GR CH (CE NE NO NW SE SW TAI) HU IN IQ IR IS IT JA JO LA LB LT MA MO MR NL NP NR NT PA PL PT RO RU SA SC SK SP SK SP SV SY SZ TR TU UK UZ YE (YEC)

Genus *Stephanopachys* Waterhouse, 1888

Stephanopachys Waterhouse, 1888: 349

Type species: *Apate substriata* Paykull, 1800: 142

†*Stephanopachys ambericus* Zahradník et Háva, 2015

Stephanopachys ambericus Zahradník et Háva, 2015: 434 [CCHH]

Distribution: P: DE (Eocene, Danish amber) RU (Eocene, Baltic amber)

Stephanopachys amplus (Casey, 1898)

Dinoderus amplus Casey, 1898: 74 [USNM]

Distribution: N: US (USL USO)

Stephanopachys asperulus (Casey, 1898)

Dinoderus asperulus Casey, 1898: 74 [USNM]

Distribution: N: US (USO)

Stephanopachys brunneus (Wollaston, 1862)

Dinoderus brunneus Wollaston, 1862: 440 [BMNH]

Distribution: P: CI

Stephanopachys conicola Fisher, 1950

Stephanopachys conicola Fisher, 1950: 48 [USNM]

Distribution: N: US (USC USO)

Stephanopachys cribratus (LeConte, 1866)

Dinoderus cribratus LeConte, 1866: 102 [MCZC]

Distribution: N: CN US (USD USE USL USN USO USP UST)

Stephanopachys densus (LeConte, 1866)

Dinoderus densus LeConte, 1866: 102 [MCZC]

Distribution: N: US (USE USL UST)

Stephanopachys dugesi Lesne, 1939

Stephanopachys dugesi Lesne, 1939b: 93 [MNHN]

Distribution: Nt: MX (MXM)

†*Stephanopachys electron* Zahradník et Háva, 2015

Stephanopachys electron Zahradník et Háva, 2015: 434 [CCHH]

Distribution: **P:** RU (Eocene, Baltic amber)

Stephanopachys himalayanus Lesne, 1932

Stephanopachys himalayanus Lesne, 1932b: 651 [MNHN IFRI]

Distribution: **I:** ID (IDP); **P:** CH (WP) ID (HP UP)

Stephanopachys hispidulus (Casey, 1898)

Dinoderus hispidulus Casey, 1898: 75 [USNM]

= *Dinoderus parvulus* Casey, 1898: 75 [USNM]

Distribution: **N:** US (USE USG UST)

Stephanopachys linearis (Kugelann, 1792)

Apate linearis Kugelann, 1792: 495 [unknown]

= *Apate elongatus* Paykull, 1800: 143 [NHRS]

Distribution: **P:** AB AR AU BY CH (NE) DE EN FI FR GE GG IN IT LA LT NR NT PL RU (CT ES FE NT WS) SC SL SK SZ UK

Stephanopachys quadricollis (Fairmaire in Marseul, 1879)

Dinoderus 4-collis Fairmaire in Marseul, 1879: 83 [MNHN]

= *Stephanopachys quadraticollis* Kocher, 1956: 114 [LC]

Distribution: **P:** AG CR FR GR ID (AP) IS IT MA MO PT SP SV_i SY TR TU UK

Stephanopachys rugosus (Olivier, 1790b)

Bostrichus rugosus Olivier, 1790b: 18 [MNHN?]

= *Dinoderus opacus* Casey, 1898: 75 [USNM]

= *Dinoderus porcatus* LeConte, 1866: 101 [MCZC]

Distribution: **A_i:** NZ; **E_i:** SF; **N:** CN (CNL CNM) US (USD USE USO UST)

Stephanopachys sobrinus (Casey, 1898)

Dinoderus sobrinus Casey, 1898: 74 [USNM]

Distribution: **N:** CN (CNN CNP) US (USA USC USO USP USU)

Stephanopachys substriatus (Paykull, 1800)

Apate substriatus Paykull, 1800: 142 [NHRS?]

= *Apate rhapontica* Motschulsky, 1853: 22 [**NN**]

= *Apate substriata* Gyllenhal in Sturm, 1826: 95 [**NN**]

= *Apate substriata* Gysselen in Dejean, 1835: 309 [MNHN?] [**NN HN**]

= *Dinoderus pacificus* Casey, 1898: 74 [USNM]

= *Rhyzopertha sachalinensis* Matsumura, 1911: 126 [EIHU]

Distribution: A_i: PW; I_i: BM IA (IAC); **N:** CN (CNA CNL CNM CNN CNP) US (USA USC USL USN USU); Nt_i: BZ HO; **P:** AL AU BH BY CR CZ EN ES FI FR GB GE GR HU IT LA LT MC MD NR PL RO RU (CT ES FE NT WS) SK SL SP SV SY SZ UK YU

†*Stephanopachys vetus* Peris, Delclos et Perrichot, 2014

Stephanopachys vetus Peris, Delclos et Perrichot, 2014: 3 [GMUR]

Distribution: **P:** FR (Cretaceous, French amber)

Subfamily *Dysidinae* Lesne, 1921

Dysididae Lesne, 1921c: 286

Type genus: *Dysides* Perty, 1832

= Apoleoninae Gardner, 1933: 3

Type genus: *Apoleon* Gorham, 1885

Genus *Apoleon* Gorham, 1885

Apoleon Gorham, 1885: 51

Type species: *Apoleon edax* Gorham, 1885

Apoleon edax Gorham, 1885

Apoleon edax Gorham, 1885: 52 [MNHN RMHN BMNH]

= *Dysides spineus* Chen et Yin, 2003: 113 [IZAS TCIS]

Distribution: **I:** BM CA IA (IAB IAJ IAS) LO MY (MYC MYS) TH VT; P_i: CH (SE)

Genus *Dysides* Perty, 1832

Dysides Perty, 1832: 113

Type species: *Dysides obscurus* Perty, 1832

Dysides obscurus Perty, 1832

Dysides obscurus Perty, 1832: 113 [ZSMC]

= *Dysides platensis* Fairmaire, 1892b: 245 [MNHN]

Distribution: **Nt:** AY (AYN) BV BZ (BZA BZS) CB FG GU PE PG VE

Subfamily *Endecatominae* LeConte, 1861

Endecatomini LeConte, 1861: 207

Type genus *Endecatomus* Mellié, 1847

= Hendecatomini Kiesenwetter, 1877: 23 [LC]

Type genus *Hendecatomus* Bach, 1852

Genus *Endecatomus* Mellié, 1847

Endecatomus Mellié, 1847: 108

Type species: *Anobium reticulatum* Herbst, 1793

= *Dictyalotus* Redtenbacher, 1847: 348

Type species: *Anobium reticulatum* Herbst, 1793

= *Hendecatomus* Bach, 1852: 108 [LC]

Endecatomus dorsalis Mellié, 1848

Endecatomus dorsalis Mellié, 1848: 218 [MNHN]

Distribution: N: US (USC USD)

Endecatomus lanatus (Lesne, 1934)

Hendecatomus lanatus Lesne, 1934a: 174 [DEIC MNHN BMNH]

Distribution: P: CH (NE) JA RU (FE)

Endecatomus reticulatus (Herbst, 1793)

Anobium reticulatum Herbst, 1793: 70 [ZMUB]

Distribution: N_i: US (UST); **P:** AU BE CZ FR FT GE HU IT NL PL RO SK SL SZ RU (ES FE)

Endecatomus rugosus (Randall, 1838)

Triphyllus rugosus Randall, 1838: 26 [unknown]

Distribution: N: CN (CNM CNP) US (USD USE USG USL USO UST)

Subfamily *Euderiinae* Lesne, 1934

Euderiitae Lesne, 1934e: 392

Type genus: *Euderia* Broun, 1880

Genus *Euderia* Broun, 1880

Euderia Broun, 1880: 344

Type species: *Euderia squamosa* Broun, 1880

Euderia squamosa Broun, 1880

Euderia squamosa Broun, 1880: 344 [MNHN]

Distribution: A: NZ

Subfamily Lyctinae Billberg, 1820

Lyctides Billberg, 1820a: 48

Type genus: *Lyctus* Fabricius, 1792

= Xylotrogidae Schönfeldt, 1887: 128

Type genus: *Xylotrogus* Stephens, 1830

Tribe *Lyctini* Billberg, 1820

Lyctides Billberg, 1820a: 48

Type genus: *Lyctus* Fabricius, 1792

Genus *Acantholyctus* Lesne, 1924

Acantholyctus Lesne, 1924: 98

Type species: *Lyctus cornifrons* Lesne, 1898

Acantholyctus cornifrons (Lesne, 1898)

Lyctus cornifrons Lesne, 1898b: 139 [MNHN]

= *Lyctus cornifrons australis* Lesne, 1914b: 335 [ZMUH]

Distribution: **E:** CD DJ EH ER ML MT MZ NB NI SF SG SO ZA; **P:** AE AG EG IN IS JO LB MO OM SA TU YE (YEC)

Acantholyctus semiermis (Lesne, 1914)

Lyctus semiermis Lesne, 1914b: 333 [SAMC]

Distribution: **E:** SF; **P:** IN

Genus *Loranthophila* Lawrence et Ślipiński, 2013

Loranthophila Lawrence et Ślipiński, 2013: 295

Type species: *Neotrichus acanthacollis* Carter et Zeck, 1937

Loranthophila acanthacollis (Carter et Zeck, 1937)

Neotrichus acanthacollis Carter et Zeck, 1937: 195 [AMSA]

= *Minthea armstrongi* Vrydagh, 1958a: 41 [ANIC? ISNB]

Distribution: **A:** AS (ASE ASQ AST)

Genus *Lycthoplites* Lesne, 1935

Lycthoplites Lesne, 1935a: 2

= *Lyctenoplus* Lesne in Colart, 1934: 242 [NN]

Type species: *Lycthoplites armatus* Lesne, 1935

Lycthoplites armatus Lesne, 1935

Lycthoplites armatus Lesne, 1935a: 2 [MRAC ISNB]

= *Lyctenoplus perarmatus* Lesne in Collart, 1934: 242 [NN]

Distribution: **E:** AA CG ZA

Genus *Lyctodon* Lesne, 1937

Lyctodon Lesne, 1937f: 319

Type species: *Lyctodon bostrychoides* Lesne, 1937

Lyctodon bostrychoides Lesne, 1937

Lyctodon bostrychoides Lesne, 1937f: 326 [MNHN]

Distribution: **A:** AS (ASE)

Genus *Lyctoxylon* Reitter, 1879

Lyctoxylon Reitter 1879: 196

Type species: *Lyctoxylon japonum* Reitter, 1879

Lyctoxylon beesonianum Lesne, 1936

Lyctoxylon beesonianum Lesne, 1936a: 133

Distribution: **P:** ID (HP UP)

Lyctoxylon convictor Lesne, 1936

Lyctoxylon convictor Lesne, 1936a: 132 [MNHN IFRI]

Distribution: **I:** ID (IDC IDS); **P:** ID (HP UP)

Lyctoxylon dentatum (Pascoe, 1866)

Minthea dentata Pascoe, 1866b: 141 [BMNH]

= *Lyctoxylon japonum* Reitter, 1879: 199 [MZPAN HNHM]

= *Lyctus seriehispidus* Kiesenwetter, 1879: 319 [ZSMC]

Distribution: **A_i:** AS (ASE) IA (SH); **E_i:** CG KY MZ SF TZ ZA ZB; **I:** BM IA (IAC IAJ) ID MY (MYC) TH VT; **N_i:** CN (CNM) US (USE USN ISO UST); **Nt_i:** PN PR; **P:** CH (CE SE SW TAI) FR_i GB_i GE_i IT_i JA NL_i PT_i

Genus *Lyctus* Fabricius, 1792

Lyctus Fabricius, 1792: 502

Type species: *Lyctus canaliculatus* Fabricius, 1792

= *Eulyctus* Jacobson, 1915: 896

Type species: *Lyctus suturalis* Faldermann, 1837

= *Lictus* Melsheimer, 1806: 8 [LC NN]

= *Xylotrogus* Stephens, 1830: 116

Type species: *Xylotrogus brunneus* Stephens, 1830

Lyctus africanus Lesne, 1907

Lyctus africanus Lesne, 1907c: 302 [MNHN MRAC ISNB]

= *Lyctus africanus capensis* Lesne, 1914b: 332 [MNHN SAMC ZMUH BMNH]

= *Lyctus africanus nigellus* Lesne, 1935a: 1 [MNHN MRAC]

= *Lyctus politus* Kraus, 1911: 118 [USNM]

Distribution: **A_i:** AS PW; **E:** BD BN CD CF CG EH GX IV MB ML MZ NI NX RW SF SG TZ SU UG ZA; **I_i:** ID PH TH; **N_i:** US (USE USL USM USN UST); **P:** AG BE_i CH_i (SE SW) CZ_i EG FR_i GB_i GE_i IN IS IT_i JA_i MO NP_i OM PA_i PL_i SA SP_i SY SZ_i TR YE (YEC YES)

Lyctus brunneus (Stephens, 1830)

Xylotrogus brunneus Stephens, 1830: 117 [BMNH]

= *Lyctus carolinae* Casey, 1891: 13 [USNM]

= *Lyctus colydioides* Dejean, 1835: 313 [MNHN?] [**NN**]

= *Lyctus costatus* Blackburn, 1888: 265 [BMNH]

= *Lyctus disputans* Walker, 1858a: 206 [BMNH]

= *Lyctus glycyrrhizae* Chevrolat in Dejean, 1835: 313 [MNHN?] [**NN**]

= *Lyctus glycyrrhizae* Chevrolat in Guérin-Méneville, 1844: 191 [**NN HN**]

= *Lyctus jatrophae* Wollaston, 1867: 112 [BMNH]

= *Lyctus parasiticus* Jacquelin du Val, 1861: 168 [**NN HN**]

= *Lyctus retractus* Walker, 1858a: 206 [BMNH]

= *Lyctus rugulosus* Montrouzier, 1861: 266 [MNHN?]

= *Silvanus retrahens* Walker, 1858a: 207 [BMNH]

= *Xylotrogus parasiticus* Stephens, 1830: 117 [**NN**]

Distribution: **A_i:** AS (ASS AST ASW) FPS (FPM) NA NZ PW SJ; **E_i:** BD BX CB CG CM CV EH EQ GO IV KY MB MU MW NX RI RW SF SWA UG ZA; **I:** BM CX IA (IAB IAJ IAN) ID MY (MYC) ID PH TH VT; **N_i:** CN (CNL CNM CNN) DEA US (USC USE USN UST); **Nt:** AY (AYS) BZ CB CU GI GL HA JC MX (MXM) PG VE; **P_i:** AF AG AU AZ BE BU BY CH (CE NO SE SW TAI) CI CR CY CZ DE EG FI FR GB GE GR IC IN IQ IR IS IT JA_i LA LB MA MD ME MO MR NL NR PL PT RU (CT NT) SB SC SK SP SV SZ TR TU UZ

Lyctus carbonarius Waltl, 1832

Lyctus carbonarius Waltl, 1832: 167 [NHMW]

= *Lyctus leacocianus* Wollaston, 1860a: 256 [BMNH]

= *Lyctus modestus* Lesne, 1911c: 534 [MNHN]

= *Lyctus planicollis* LeConte, 1858: 74 [MCZC]

Distribution: **A_i:** AS (ASE ASQ AST ASW) NZ; **E_i:** SF; **I_i:** IA (IAJ); **N:** CN (CNLCNM CNN) MX (MXC) US (USC USD USE USG USI USN USO UST); **Nt_i:** AY (AYS) CB GT MX (MXM) PN; **P_i:** AU BE EG FI FR GB GE IS IT JA MR NL SV SZ

Lyctus caribeanus Lesne, 1931

Lyctus caribeanus Lesne, 1931a: 96 [DEIC BMNH MNHN USNM]

Distribution: **N:** US_i (UST); **Nt:** DO DR GL GT HA MNT MX (MXT) PN PG PR

Lyctus cavicollis LeConte, 1866

Lyctus cavicollis LeConte, 1866: 103 [MCZC]

Distribution: A_i: AS (AST); **N:** CN (CNN) US (USC USD USE); **Nt:** MX; P_i: AU BE FR GB GE IN LU NL SZ

Lyctus chacoensis Santoro, 1960a

Lyctus chacoensis Santoro, 1960a: 101 [MACN MLPA ISNB]

Distribution: **Nt:** AY (AYN)

Lyctus cinereus Blanchard, 1851

Lyctus cinereus Blanchard, 1851: 438 [MNHN]

= *Lyctus argentinensis* Santoro, 1960b: 187 [MANC]

= *Lyctus chilensis* Gerberg, 1957: 21 [USNM]

= *Lyctus nitidicollis* Reitter, 1879: 197 [HNHM]

= *Lyctus patagonicus* Santoro, 1960c: 191 [MLPA]

Distribution: **Nt:** AY (AYN AYS) BZ (BZS) CL (CLN) CB

Lyctus discedens Blackburn, 1888

Lyctus discedens Blackburn, 1888: 267 [BMNH]

= *Lyctus malayanus* Lesne, 1910a: 254 [MNHN]

Distribution: **A:** AS (ASE ASN ASS) IA (IAN) NZ PW; **I:** CX IA (IAS) ID (IDS) MY (MYC); P_i: GE

Lyctus hipposideros Lesne, 1908

Lyctus hipposideros Lesne, 1908c: 356 [MNHN]

Distribution: **E:** CD CG ER GA KY ML MT MZ NI NX SF SG SU TZ ZA ZB ZI; **P:** AG FI_i GE_i IT_i MO SZ_i

Lyctus kosciuszkoi Borowski et Węgrzynowicz, 2007

Lyctus kosciuszkoi Borowski et Węgrzynowicz, 2007a: 18

= *Lyctus parvulus* Casey, 1885: 175 [USNM MCZC] [**HN**]

Distribution: **N:** US (USC USO)

Lyctus linearis (Goeze, 1777)

Dermestes linearis Goeze, 1777: 148 [unknown]

= *Dermestes oblongus* Geoffroy in Fourcroy, 1785: 19 [unknown]

= *Dermestoides unipunctatus* Herbst, 1783: 40 [ZMUB]

= *Lictus axillaris* Melsheimer, 1806: 9 [?-MCZC] [**NN**]

= *Lictus canaliculatus* Melsheimer, 1806: 9 [?-MCZC] [**NN**]

= *Lictus striatus* Melsheimer, 1806: 9 [?-MCZC] [**NN**]

= *Lyctus axillaris* Melsheimer, 1846: 113 [MCZC]

= *Lyctus canaliculatus* Fabricius, 1792: 504 [ZMUC] [**NN HN**]

= *Lyctus canaliculatus* Strum, 1826: 166 **P:** GE [unknown] [**NN HN**]

= *Lyctus canaliculatus nitidus* Dahl in Dejean, 1835: 313 [MNHN?] [**NN**]

= *Lyctus duftschmidi* Gozis, 1881: 135 [**RN**]
= *Lyctus fuscus* Paykull, 1800: 332 [NHRS]
= *Lyctus fuscus* morpha *crassicollis* Lesne, 1916: 96 [**IN**]
= *Lyctus oblongus* Latreille in A. Villa et G. B. Villa, 1844: 465 [**NN**]
= *Lyctus oblongus pusillus* Stephens, 1830: 118 [BMNH]
= *Lyctus pubescens* Duftschmid, 1825: 148 [OLML] [**HN**]
= *Lyctus striatus* Melsheimer, 1846: 112 [MCZC] [**NN HN**]

Distribution: **A_i:** AS (ASE ASQ AST) NZ; **E_i:** LI; **N:** CN (CNL CNM CNN) US (USC USD USG USL USN USO UST); **Nt_i:** AY (AYS) CB CL; **P:** AB AG AL AU AZ BE BH BU BY CH (CE NO SE) CI CR CY CZ DE EG FI FR GB GE GG GR HU IC IN IR IT JA KZ LA LT LU MC MD ME MO NL NR PL PT RO RU (CT ES NT ST) SB SK SL SP SV SY SZ TM TR TU UK UZ YU

Lyctus longicornis Reitter, 1879

Lyctus longicornis Reitter, 1879: 197 [HNHM]

Distribution: **Nt:** CB

Lyctus opaculus LeConte, 1866

Lyctus opaculus LeConte, 1866: 103 [MCZC]
= *Lyctus brevipennis* Casey, 1924: 183 [USNM]

Distribution: **N:** CN (CNL) US (USD USE USG UST USU)

Lyctus parallelocollis Blackburn, 1888

Lyctus parallelocollis Blackburn, 1888: 266 [BMNH]

Distribution: **A:** AS (ASE ASS ASQ) PW; **E_i:** SF; **Nt_i:** CC; **P_i:** IS

Lyctus pubescens Panzer, 1793

Lyctus pubescens Panzer, 1793a: 17 [ZMUB]
= *Lyctus bicolor* Comolli, 1837: 41 [unknown]
= *Lyctus bicolor* Perroud in A. Villa et G. B. Villa, 1844: 465 [unknown] [**NN HN**]
= *Lyctus bicolor* Perroud in Sturm, 1843: 234 [unknown] [**NN HN**]
= *Lyctus caucasicus* Tournier, 1874: 412 [MNHN]
= *Lyctus subarmatus* Megerle in Dejean, 1835: 313 [MNHN?] [**NN**]

Distribution: **A_i:** AS NZ PW; **N_i:** CN; **P:** AB AL AR AU BE BH BU CH (TAI) CR CY CZ FR GB GE GR HU IN IT KZ LS LU MC MD NL PL PT RO RU (CT ST) SK SL SP SZ TR UK YU

Lyctus simplex Reitter, 1879

Lyctus simplex Reitter, 1879: 198 [HNHM? MNHN?]

Distribution: **E_i:** SF; **Nt:** AY (AYN) BV CB EC PE PG; **P_i:** GE SV

Lyctus sinensis Lesne, 1911

Lyctus sinensis Lesne, 1911a: 48 [MNHN]

Distribution: **A_i:** AS (ASE) NZ PW; **P:** CH (NE NO SE SW TAI) CZ_i GB_i GE_i JA SC_i SP_i SZ_i

Lyctus suturalis Faldermann, 1837

Lyctus suturalis Faldermann, 1837: 255 [ZMUM?]

= *Lyctus deyrollei* Tournier, 1874: 411 [MNHN]

= *Lyctus shestakovi* Fursov, 1939: 89 [ZMUM]

= *Lyctus turkestanicus* Fursov, 1939: 89 [ZMUM] [**HN**]

Distribution: P: AB AR BY CH (NW) GG ID (KA) IN KA KZ MD RU (ST) TD TM UK UZ

Lyctus tomentosus Reitter, 1879

Lyctus tomentosus Reitter, 1879: 198 [HNHM]

= *Lyctus griseus* Gorham, 1883: 212 [BMNH]

Distribution: $\mathbf{A_i}$: HI; $\mathbf{I_i}$: PH TH; **N:** US; **Nt:** GT MX

Lyctus turkestanicus Lesne, 1935

Lyctus turkestanicus Lesne, 1935c: 300 [MNHN]

= *Lyctus asiaticus* Iablokoff-Khnzorian, 1976: 96 [unknown]

Distribution: P: CH (NW) KI TD TM UZ

Lyctus villosus Lesne, 1911

Lyctus villosus Lesne, 1911c: 537 [MNHN]

Distribution: $\mathbf{A_i}$: HI; $\mathbf{N_i}$: US (USE USO); **Nt:** „Antiles" CB CU EL MX (MXM MXW) PR SM

Lictus Melsheimer, 1806, other nomina nuda

Lictus americanus Melsheimer, 1806: 9 [MCZC] [**NN**]

Lictus fuliginosus Melsheimer, 1806: 9 [MCZC] [**NN**]

Lictus fuscus Melsheimer, 1806: 9 [MCZC] [**NN**]

Lictus glaber Melsheimer, 1806: 9 [MCZC] [**NN**]

Lictus histeroides Melsheimer, 1806: 9 [MCZC] [**NN**]

Lictus impressus Melsheimer, 1806: 9 [MCZC] [**NN**]

Lictus marginalis Melsheimer, 1806: 9 [MCZC] [**NN**]

Lictus minutus Melsheimer, 1806: 10 [MCZC] [**NN**]

Lictus politus Kugelann in Melsheimer, 1806: 9 [MCZC] [**NN**]

Lictus porcatus Melsheimer, 1806: 9 [MCZC] [**NN**]

Lictus punctatus Melsheimer, 1806: 9 [MCZC] [**NN**]

Lictus sulcatus Melsheimer, 1806: 9 [MCZC] [**NN**]

Lictus unicolor Melsheimer, 1806: 10 [MCZC] [**NN**]

Lyctus Fabricius, 1792, other nomina nuda

Lyctus americanus Dejean, 1835: 313 [MNHN?] [**NN**]

Lyctus angulatus Sturm, 1843: 234 **P:** LT [unknown] [**NN**]

Lyctus destructor Dejean, 1835: 313 [MNHN?] [**NN**]

Lyctus punctatus Strum, 1826: 166 **P:** GE [unknown] [**NN**]

Lyctus punctatus Strum, 1843: 234 **N:** America borealis [unknown] [**NN HN**]

Lyctus rhabarbari Boudier in Dejean 1835: 313 [MNHN?] [NN]
Lyctus sulcatus Strum, 1826: 166 N: „America borealis“ [unknown] [NN]

Genus *Minthea* Pascoe, 1866

Minthea Pascoe 1866a: 97
Type species: *Minthea squamigera* Pascoe, 1866
= *Eulachus* Blackburn in Blackburn et Sharp, 1885: 141
Type species: *Eulachus hispidus* Blackburn in Blackburn et Sharp, 1885
= *Lyctopholis* Reitter, 1879: 199
Type species: *Lyctopholis stichothrix* Reitter, 1879

Minthea apicata Lesne, 1935

Minthea apicata Lesne, 1935a: 1 [MNHN MRAC]
Distribution: E: GH KY RW ZA

Minthea bivestita Lesne, 1937

Minthea bivestita Lesne, 1937f: 320 [MNHN]
Distribution: I: ID VT

Minthea humericosta Lesne, 1936

Minthea humericosta Lesne, 1936a: 131 [BMNH]
Distribution: A: „Oceania“ PW; **I:** MY (MYC MYS) PH TH VT

Minthea obsita (Wollaston, 1867)

Lyctus obsitus Wollaston, 1867: 112 [BMNH]
Distribution: E: BD CD CG CF CV EH GA GH GO GX MB ML MZ NI NX RW SF SG TG TZ ZA ZB ZI; **N_i:** US (USD USE); **$Nt_{i:}$** CU „South America“; **P_i:** AU IT PT

Minthea reticulata Lesne, 1931

Minthea reticulata Lesne, 1931a: 98 [MNHN BMNH USNM]
Distribution: A: AS (ASE) IA (IAN SH) STI PW; **I:** CX IA (IAB IAC IAJ IAS) MY (MYS) PH SX TH VT; **N_i:** US (USG); **P:** CH (SE TAI) GB_i

Minthea rugicollis (Walker, 1858)

Ditoma rugicollis Walker, 1858a: 206 [BMNH]
= *Eulachus hispidus* Blackburn in Blackburn et Sharp, 1885: 141 [BMNH]
= *Lyctopholis foveicollis* Reitter, 1879: 199 [HNHM]
= *Minthea similata* Pascoe, 1866b: 141 [BMNH]
Distribution: A_i: AS (ASE) FJ HI IA (IAN SH) NA PW SJ; **E_i:** AA BN CF CG CV EH GH GX IV KY MB MU MZ NX RI RW SF SU TZ UG ZA; **I:** AI BM CHI CX IA (IAJ IAS) ID LO MY (MYC MYS) PH TH

VT; N_i: US (USE USG USO UST); Nt_i: AY BZ CB CL CU DR EC ES FG GI GL GU HA MTS MX PG PN PR; P_i: AU BE CH (CE SE SW TAI) DE FI FR GB GE IS IT JA NL SV TD

Minthea squamigera Pascoe, 1866

Minthea squamigera Pascoe, 1866a: 97 [BMNH]

= *Lyctopholis stichothrix* Reitter, 1879: 199 [HNHM]

Distribution: **E_i:** GU; **N:** US (UST); **Nt:** AY BZ CB GU PE PG SM; **P_i:** AU GB GR

Tribe *Cephalotomini* Liu in Liu et Schönitzer, 2011

Cephalotomini Liu in Liu et Schönitzer, 2011: 120

Type genus: *Cephalotoma* Lesne, 1911

Genus *Cephalotoma* Lesne, 1911

Cephalotoma Lesne, 1911b: 204

Type species: *Cephalotoma singularis* Lesne, 1911

= *Lyctoderma* Lesne, 1911b: 204

Type species: *Tristaria africana* Grouvelle, 1900

Cephalotoma africanum (Grouvelle, 1900)

Tristaria africana Grouvelle, 1900: 424 [ISNB]

Distribution: **E:** CG CK GX ZA

Cephalotoma ambiguum (Lesne, 1936)

Lyctoderma ambiguum Lesne, 1936a: 133 [MNHN IFRI]

Distribution: **I:** ID (IDC IDS)

Cephalotoma coomani (Lesne, 1932)

Lyctoderma coomani Lesne, 1932d: 622 [MNHN]

Distribution: **I:** TH VT; **P:** CH (CE)

Cephalotoma perdepressa Lesne, 1937

Cephalotoma perdepressa Lesne, 1937f: 323 [MNHN]

Distribution: **I:** MY (MYC) PH TH VT; **P:** CH (TAI)

Cephalotoma singularis Lesne, 1911

Cephalotoma singularis Lesne, 1911b: 207 [MNHN]

= *Cephalotoma tonkinea* Lesne, 1932d: 623 [MNHN]

Distribution: **A:** IA (IAN) PW; **I:** IA (IAS) TH VT; **P_i:** FR

Cephalotoma testaceum (Lesne, 1913)

Lyctoderma testaceum Lesne, 1913c: 563 [MRAC]

Distribution: E: CG GO MZ ZA

Tribe *Trogoxylini* Lesne, 1921

Trogoxylini Lesne, 1921b: 231

Type genus: *Trogoxylon* LeConte, 1862

= Tristariini Lesne, 1921c: 287

Type genus: *Tristaria* Reitter, 1878

= Trogoxylini Liu & Schönitzer, 2011 (sic!) = Trogoxylini Liu, Leavenggod et Bernal, 2022: 32 [HN]

Genus *Lyctopsis* Lesne, 1911

Lyctopsis Lesne, 1911b: 204

Type species: *Lyctopsis pachymera* Lesne, 1911

Lyctopsis inquilina Lesne, 1932

Lyctopsis inquilina Lesne, 1932d: 626 [MNHN]

Distribution: E: MZ

Lyctopsis pachymera Lesne, 1911

Lyctopsis pachymera Lesne, 1911b: 205 [MNHN]

Distribution: E: CD ZA; P: AE

Lyctopsis scabricollis Lesne, 1911

Lyctopsis scabricollis Lesne, 1911b: 205 [MNHN]

Distribution: A_i: AS; E: DJ SO; P: OM SA

Genus *Phyllyctus* Lesne, 1911

Phylyctus Lesne, 1911b: 204

Type species: *Tristaria gounellei* Grouvelle, 1896

Phyllyctus gounellei (Grouvelle, 1896)

Tristaria gounellei Grouvelle, 1896: 193 [MNHN]

Distribution: Nt: AY (AYN) BZ (BZC BZE) PG

Genus *Tristaria* Reitter, 1878

Tristaria Reitter, 1878: 320

Type species: *Tristaria grouvellei* Reitter, 1878

Tristaria grouvellei Reitter, 1878

Tristaria grouvellei Reitter, 1878: 321 [MNHN]

= *Tristaria fulvipes* Reitter, 1878: 322 [MNHN]

= *Tristaria labralis* Blackburn, 1892: 29 [BMNH]

Distribution: A: AS (ASE ASQ ASS ASW)

Genus *Trogoxylon* LeConte, 1862

Trogoxylon LeConte, 1862: 209

Type species: *Xylotrogus parallelipipedus* Melsheimer, 1846

Trogoxylon aequale (Wollaston, 1867)

Lyctus aequale Wollaston, 1867: 111 [BMNH]

= *Lyctus californicus* Casey, 1891: 14 [USNM]

= *Lyctus curtulus* Casey, 1891: 15 [USNM]

Distribution: $\mathbf{A_i}$: HI; $\mathbf{E_i}$: CG CK CV GX NX ZA; $\mathbf{I_i}$: PH; **N:** MX (MXC) US (USD USL USO UST); **Nt:** BI BZ (BZE) CC CU DR FG GT MX (MXY) PR; $\mathbf{P_i}$: BE NL PT SZ

Trogoxylon angulicollis Santoro, 1960

Trogoxylon angulicollis Santoro, 1960c: 194 [MLPA MANC ISNB]

Distribution: Nt: AY (AYS)

Trogoxylon auriculatum Lesne, 1932

Trogoxylon auriculatum Lesne, 1932b: 654 [IFRI MNHN]

Distribution: I: ID (IDP) TH; **P:** ID (HP UP) PA

Trogoxylon caseyi Lesne, 1937

Trogoxylon caseyi Lesne, 1937c: 240 [USNM] (**RN**)

= *Lyctus rectangulum* Casey, 1924: 184 [USNM] [**HN**]

Distribution: N: US (USO); **Nt:** DR

Trogoxylon giacobbii Santoro, 1957

Trogoxylon giacobbii Santoro, 1957a: 153 [unknown]

Distribution: Nt: AY (AYN) BZ PG

Trogoxylon impressum (Comolli, 1837)

Lyctus impressum Comolli, 1837: 40 [unknown]

= *Lyctus castaneus* Perroud in Comolli, 1837: 40 [MNHN?] [**NN**]

= *Lyctus glabratus* Ullrich in A. Villa et G. B. Villa, 1844: 465 [unknown] [**NN HN**]

= *Lyctus glabratus* A. Villa et G. B. Villa in Dejean, 1843: 234 [unknown] [**NN**]

= *Lyctus impressum* Dejean, 1821: 103 [MNHN?] [NN]

= *Lyctus impressum* Dejean, 1835: 313 [MNHN?] [NN HN]

= *Lyctus impressus capitalis* Schaufuss, 1882: 534 [unknown]

= *Lyctus laevis* Comolli, 1837: 40 [NN]

= *Lyctus laevis* Galeazzi, 1854: 15 [NN HN]

= *Lyctus laevis* Galeazzi in A. Villa et G. B. Villa, 1844: 465 [NN HN]

= *Lyctus laevipennis* Faldermann, 1837: 256 [ZMUM?]

Distribution: **A_i:** AS NZ; **E_i:** SF; **N_i:** CN (CNL CNM) US (USD USE USG USO UST); **Nt_i:** AY (AYN AYS) CL MX; **P:** AG AL AR AU_i BE BH BU CI CH CR CY CZ DE_i EG FI_i FR GB GE_i GG GR HU IN IQ IT JA LB LU MA MC MO NL NR_i PT RO RU (ST) SK SL SP SV_i SY SZ TM TR TU UK YU

Trogoxylon parallelipipedum (Melsheimer, 1846)

Xylotrogus parallelipipedum Melsheimer, 1846: 112 [MCZC]

= *Lictus parallellipipedus* Melsheimer, 1806: 10 [MCZC] [NN]

Distribution: **A_i:** AS (ASE) FJ; **E_i:** TZ; **N:** CN (CNL CNM) US (USD USE USO UST); **Nt_i:** EC (GI) HA; **P_i:** AU GB HU IT PT

Trogoxylon praeustum (Erichson, 1847)

Lyctus praeustus Erichson, 1847: 88 [ZMUB]

= *Lyctus prostomoides* Gorham, 1883: 212 [BMNH ISNB]

Distribution: **I_i:** PH; **N:** US (USE USO UST); **Nt:** AY (AYN AYS) BV BZ CC GN GT MX (MXE MXM MXT MXV) NG PE PN STV; **P_i:** FI GE SZ

Trogoxylon punctatum LeConte, 1866

Trogoxylon punctatum LeConte, 1866: 104 [MCZC]

Distribution: **N:** MX (MXC) US (UST); **Nt:** MX (MXV)

Trogoxylon punctipenne (Fauvel, 1904)

Lyctus punctipennis Fauvel, 1904: 155 [ISNB]

Distribution: **A:** AS (ASE ASQ) IA (IAN) NA PW; **I:** ID IA (IAS) MY (MYC) TH

Trogoxylon rectangulum Lesne, 1921

Trogoxylon rectangulum Lesne, 1921b: 229 [MNHN]

Distribution: **Nt:** DR HA

Trogoxylon recticolle Reitter, 1879

Trogoxylon recticolle Reitter, 1879: 199 [HNHM]

= *Trogoxylon ingae* Santoro, 1956: 46 [ININ ISNB]

Distribution: **Nt:** AY (AYN AYS) CL (CLN)

Trogoxylon spinifrons (Lesne, 1910)

Lyctus spinifrons Lesne, 1910b: 303 [MNHN BMNH]

Distribution: **A:** IA (IAN) PW; **I:** ID (IDB) TH VT

Trogoxylon ypsilon Lesne, 1937

Trogoxylon ypsilon Lesne, 1937f: 322 [ANIC MNHN]

Distribution: **A:** AS (ASE ASQ ASS AST) PW SS

Genus *Trogoxylyctus* Węgrzynowicz et Borowski, 2015

Trogoxylyctus Węgrzynowicz et Borowski, 2015b: 58

Type species: *Trogoxylyctus australiensis* Węgrzynowicz et Borowski, 2015

Trogoxylyctus australiensis Węgrzynowicz et Borowski, 2015

Trogoxylyctus australiensis Węgrzynowicz et Borowski, 2015b: 59 [HNHM]

Distribution: **A:** AS (ASN)

Subfamily *Polycaoninae* Lesne, 1896

Polycaoninae Lesne, 1896a: 96

Type genus: *Polycaon* Castelnau, 1836

†Genus *Cretolgus* Legalov et Háva, 2020

Cretolgus Legalov et Háva, 2020: 2

Type species: *Cretolgus minimus* Legalov et Háva, 2020

†*Cretolgus minimus* Legalov et Háva, 2020

Cretolgus minimus Legalov et Háva, 2020: 2 [ISEA]

Distribution: **I:** BM (Cretaceous, Burmese amber)

Genus *Melalgus* Dejean, 1835

Melalgus Dejean, 1835: 309

Type species: *Apate femoralis* Fabricius, 1792

= *Exopioides* Guérin-Méneville, 1844: 187

Type species: *Exopioides carinatus* Guérin-Méneville, 1844

= *Exopsoides* Guérin-Méneville, 1845: 17 [LC]

= *Heterarthron* Guérin-Méneville, 1844: 186

Type species: *Apate femoralis* Fabricius, 1792

= *Heterarthron* Guérin-Méneville in Dejean, 1835: 334 [MNHN?] [**NN**]

Melalgus amoenus (Lesne, 1911)

Heterarthron amoenum Lesne, 1911d: 46 [BMNH]

Distribution: **N:** US (USO); **Nt:** BZ CB MX

Melalgus batillus (Lesne, 1902)

Heterarthron batillum Lesne, 1902b: 223 [MNHN]

= *Melalgus japonicus* Chûjô, 1973: 10 [EUMJ]

= *Heterarthron talpula* Lesne, 1911d: 47 [MNHN]

Distribution: **I:** ID (IDE) VT; **P:** CH (CE HAI SE SW) JA

Melalgus borneensis (Lesne, 1911)

Heterarthron borneense Lesne, 1911d: 45 [MNHN]

Distribution: **I:** MY (MYC MYS)

Melalgus caribeanus (Lesne, 1906)

Heterarthron caribeanum Lesne, 1906a: 399 [MNHN BMNH]

Distribution: **Nt:** GL MNT SN STV TT VE

Melalgus confertus (LeConte, 1866)

Polycaon confertus LeConte, 1866: 103 [ICCM]

= *Exopioides incisa* LeConte, 1868: 64 [MCZC]

Distribution: **N:** CN (CNN) US (USC USN)

Melalgus crassulus (Lesne, 1911)

Heterarthron crassulum Lesne, 1911d: 46 [BMNH]

Distribution: **Nt:** MX (MXT)

Melalgus digueti (Lesne, 1911)

Heterarthron digueti Lesne, 1911d: 47 [MNHN]

Distribution: **Nt:** MX (MXT)

Melalgus exesus (LeConte, 1858)

Exops exesus LeConte, 1858: 74 [MCZC]

Distribution: **N:** MX (MXC) US (USC USO); **Nt:** „Antilles“ CC GT HO MX (MXM MXT MXV) NG PN

Melalgus feanus (Lesne, 1899)

Heterarthron feanus Lesne, 1899d: 634 [MCSN]

Distribution: **I:** BM ID (IDE) TH VT; **P:** CH (CE SW) ID (UP)

Melalgus femoralis (Fabricius, 1792)

Apate femoralis Fabricius, 1792: 361 [ZMUC]

= *Apate gonagra* Fabricius, 1798: 156 [ZMUC?]

Distribution: N: US (USE USO); **Nt:** CC CU DO DR FG GL GN GT GU HA JC LW MNT MX PR STB STT STI STV TT VI

Melalgus gracilipes (Blanchard in Blanchard et Brullé, 1843)

Psoa gracilipes Blanchard in Blanchard et Brullé, 1843: 205 [MNHN]

= *Exopioides carinatus* Guérin-Méneville, 1844: 187 [MNHN?]

= *Melalgus cylindrica* Dejean, 1835: 309 [MNHN?] [NN]

Distribution: Nt: AY (AYN) BV BZ (BZC BEZ BZS) CB

Melalgus jamaicensis (Lesne, 1906)

Heterarthron jamaicense Lesne, 1906a: 397 [MNHN]

Distribution: Nt: BI CMS JC

Melalgus longitarsus (Lesne, 1911)

Heterarthron longitarse Lesne, 1911d: 46 [BMNH]

Distribution: Nt: MX

Melalgus megalops (Fall, 1901)

Polycaon megalops Fall, 1901: 254 [CASC]

Distribution: N: US (USC)

Melalgus parvidens (Lesne, 1895)

Heterarthron parvidens Lesne, 1895a: 169 [MNHN]

Distribution: Nt: BZ (BZC BZE BZM BZS)

Melalgus parvulus (Lesne, 1925)

Heterarthron parvulum Lesne, 1925a: 29 [DIEC]

Distribution: Nt: MX

Melalgus plesiobatillus Liu et Beaver, 2023a

Melalgus plesiobatillus Liu et Beaver, 2023a: 275 [NMEG]

Distribution: P: CH (CE)

Melalgus plicatus (LeConte, 1874)

Polycaon plicatus LeConte, 1874: 65 [MCZC]

= *Polycaon obliquus* LeConte, 1874: 66 [MCZC]

Distribution: N: US (USE USO); **Nt:** BZ CB CU JC MX (MXS MXT)

Melalgus rufipes (Blanchard in Blanchard et Brullé, 1843)
Psoa rufipes Blanchard in Blanchard et Brullé, 1843: 205 [MNHN]
Distribution: Nt: AY BV PG

Melalgus strigipennis (Lesne, 1937)
Heterarthron strigipenne Lesne, 1937f: 323 [MNHN RMNH DEIC]
Distribution: N: US (USO); **Nt:** BZ CB MX (MXM MXS)

Melalgus subdepressus (Lesne, 1897)
Heterarthron subdepressus Lesne, 1897c: 146 [MNHN BMNH]
Distribution: Nt: MX (MXM MXT)

Melalgus truncatus (Guerin-Méneville, 1844)
Heterarthron truncatum Guerin-Méneville, 1844: 186 [MNHN?]
Distribution: Nt: GU VE

Melalgus valleculatus (Lesne, 1913)
Heterarthron valleculatum Lesne, 1913a: 191 [MACN]
Distribution: Nt: AY (AYN)

Melalgus Dejean, 1835, other nomina nuda

Melalgus elongata Sturm, 1843: 231 **Nt:** CB [unknown] [**NN**]
Melalgus femoralis Olivier in Dejean, 1835: 309 **Nt:** DR [MNHN?] [**NN**]

Genus *Polycaon* Castelnau, 1836

Polycaon Castelnau, 1836: 30
Type species: *Polycaon chilensis* Castelnau, 1836: 30
= *Alloeocnemis* LeConte, 1853: 233
Type species: *Alloeocnemis stoutii* LeConte, 1853
= *Exops* Curtis, 1839: 203
Type species: *Exops bevani* Curtis, 1839

Polycaon chilensis (Erichson, 1834)
Psoa chilensis Erichson, 1834: 266 [ZMUB]
= *Exops bevani* Curtis, 1839: 204 [BMNH]
= *Melalgus chilensis* Lacordaire in Dejean, 1835: 309 [MNHN?] [**NN**]
= *Polycaon chiliensis* Castelnau, 1836: 30 [MNHN] [**HN**]
Distribution: Nt: AY (AYN AYS) BV BZ CL (CLN) PE

Polycaon granulatus Van Dyke, 1923

Polycaon granulatus Van Dyke, 1923: 43 [CASC]

Distribution: N: US (USC)

Polycaon punctatus LeConte, 1866

Polycaon punctatus LeConte, 1866: 102 [ICCM]

= *Polycaon pubescens* LeConte, 1866: 102 [MCZC]

Distribution: N: US (USC)

Polycaon sinensis Liu et Beaver, 2023a

Polycaon sinensis Lie et Beaver, 2023a: 273 [NMEG]

Distribution: P: CH (CE)

Polycaon stoutii (LeConte, 1853)

Alloeocnemis stoutii LeConte, 1853: 233 [MCZC]

= *Exops ovicollis* LeConte, 1859: 49 [MCZC]

Distribution: N: CN (CNN) US (USC USE USN USO USU); **Nt:** MX (MXS); **P_i:** IT PL

Subfamily *Psoinae* Blanchard, 1851

Psoitas Blanchard, 1851: 434

Type genus: *Psoa* Herbst, 1797

= Psoites Jacquelin du Val, 1861: 232 [**HN**]

Tribe *Chileniini* Lesne, 1921

Chileniidae Lesne, 1921c: 287

Type genus: *Chilenius* Lesne, 1921

Genus *Chilenius* Lesne, 1921

Chilenius Lesne, 1921c: 287

Type species: *Exopioides spinicollis* Fairmaire et Germain, 1861

Chilenius spinicollis (Fairmaire et Germain, 1861)

Exopioides spinicollis Fairmaire et Germain, 1861: 4 [MNHN]

Distribution: Nt: CL (CLN)

Chilenius tabulifrons Lesne, 1935

Chilenius tabulifrons Lesne, 1935b: 34 [MNHN CASC ISNB]

Distribution: Nt: CL (CLN)

Tribe *Psoini* Blanchard, 1851

Psoitas Blanchard, 1851: 434

Type genus: *Psoa* Herbst, 1797

Genus *Coccographis* Lesne, 1901

Coccographis Lesne, 1901b: 349

Type species: *Coccographis nigrorubra* Lesne, 1901

Coccographis nigrorubra Lesne, 1901

Coccographis nigrorubra Lesne, 1901b: 349 [MNHN BMNH]

Distribution: I: LO VT; **P:** CH (SW)

Genus *Heteropsoa* Lesne, 1895

Heteropsoa Lesne, 1895a: 169

Type species: *Heteropsoa australis* Lesne, 1895

Heteropsoa australis Lesne, 1895

Heteropsoa australis Lesne, 1895a: 169 [MNHN]

Distribution: E: SF

Heteropsoa macrops Lesne, 1938

Heteropsoa macrops Lesne, 1938b: 173 [BMNH]

Distribution: E: SF

Genus *Psoa* Herbst, 1797

Psoa Herbst, 1797: 214

Type species: *Psoa vienensis* Herbst, 1797

= *Acrepis* LeConte, 1852: 213

Type species: *Acrepis maculata* LeConte, 1852

Psoa dubia (Rossi, 1792)

Dermestes dubius Rossi, 1792: 17 [ZMUB?]

= *Attelabus sanguineus* Giorna, 1792: 48 [unknown]

= *Psoa herbsti* Küster, 1847: no. 45 [unknown]

= *Psoa italica* Dejean, 1821: 101 [MNHN?] [**NN**]

= *Psoa italica* Dejean, 1835: 309 [MNHN?] [**NN HN**]

= *Psoa italica* Serville in Latreille, Le Peletier de Saint-Fargeau, Serville et Guérin-Méneville, 1825: 224 [unknown] [**HN**]

Distribution: P: AG AL AU BH BU CR CY EG FR GE GR HU IT LB LE MO PT SL SP TR TU UK YU

Psoa maculata (LeConte, 1852)

Acrepis maculata LeConte, 1852: 213 [unknown]

= *Psoa cleroides* Lesne, 1913b: 273 [DEIC]

Distribution: N: US (USC)

Psoa quadrinotata Blanchard, 1851

Psoa quadrinotata Blanchard, 1851: 436 [MNHN]

Distribution: Nt: CL

Psoa quadrisignata (Horn, 1868)

Acrepis 4-signata Horn, 1868: 135 [ICCM]

= *Psoa sexguttata* Lesne, 1906a: 393 [MNHN]

Distribution: I: MY; **N:** CN (CNN) US (USC USN); **Nt:** MX

Psoa rubripennis Park in Park, Lee et Hong, 2015

Psoa rubripennis Park in Park, Lee et Hong, 2015: 300 [RIFID]

Distribution: P: SC

Psoa viennensis Herbst, 1797

Psoa viennensis Herbst, 1797: 215 [ZMUB]

= *Psoa grandis* Motschulsky, 1845: 92 [ZMUM]

= *Psoa viennensis* Dejean, 1821: 101 [MNHN?] [**NN**]

= *Psoa viennensis* Fabricius in Dejean, 1835: 309 [MNHN?] [**NN HN**]

= *Psoa viennensis* Fabricius in Strurm, 1843: 231 [hnknown] [**NN HN**]

= *Psoa viennensis* Strurm, 1826: 188 [unknown] [**NN HN**]

Distribution: P: AB AL AR AU BH BU CR CY CZ GE GG GR HU IN IT KZ MC MD RO RU (ST) SK SL TR UK YU

Genus *Psoidia* Lesne, 1912

Psoidia Lesne, 1912a: 377

Type species: *Psoidia pexicollis* Lesne, 1912a

Psoidia pexicollis Lesne, 1912

Psoidia pexicollis Lesne, 1912a: 377 [BMNH]

Distribution: I: ID (IDW)

Genus *Sawianus* Zahradník et Háva, 2016

Sawianus Zahradník et Háva, 2016: 298

Type species: *Sawianus ornatus* Zahradník et Háva, 2016

Sawianus ornatus Zahradník et Háva, 2016

Sawianus ornatus Zahradník et Háva, 2016: 299 [JHAC]

Distribution: **I:** TH

Genus *Stenomera* Lucas, 1850

Stenomera Lucas, 1850: 38

Type species: *Stenomera blanchardii* Lucas, 1850

Stenomera assyria Lesne, 1895

Stenomera assyria Lesne, 1895a: 169 [MNHN]

Distribution: **P:** CY IN IQ SY TR

Stenomera blanchardii Lucas, 1850

Stenomera blanchardii Lucas, 1850: 41 [MNHN]

Distribution: **P:** AG MO TR TU

List of Species Not Belonging to Bostrichidae

Scientific Name

Apate Fabricius, 1775

asperata Sturm, 1843: 231 [**NN**]	Curculionidae, Scolytinae [*Cryphalus saltuarius* (Weise, 1891)]
bivittata Kirby, 1837: 192	Curculionidae, Scolytinae [*Trypodendron lineatum* (Olivier, 1790)]
brevicornis Kirby, 1837: 194	Curculionidae, Scolytinae [*Polygraphus rufipennis* (Kirby, 1837)]
dispar Fabricius, 1792: 363	Curculionidae, Scolytinae [*Anisandrus dispar* (Fabricius, 1792)]
fagi Fabricius, 1798: 157	Curculionidae, Scolytinae [*Ernoporicus fagi* (Fabricius, 1798)]
fronticornis Panzer, 1805: 7	Ciidae [*Sulcacis fronticornis* (Panzer, 1809)]
inurbanus Broun, 1880: 126	Curculionidae, Scolytinae [*Mesoscolytus inurbanus* (Broun, 1904)]
limbatus Fabricius, 1787: 33	Curculionidae, Scolytinae [*Trypodendron domesticum* (Linnaeus, 1758)]
nigriceps Kirby, 1837: 193	Curculionidae, Scolytinae [*Polygraphus rufipennis* (Kirby, 1837)]

rufipennis Kirby, 1837: 193	Curculionidae, Scolytinae [*Polygraphus rufipennis* (Kirby, 1837)]
rufitarsis Kirby, 1837: 193	Curculionidae, Scolytinae [*Trypodendron rufitarsis* (Kirby, 1837)]
signata Fabricius, 1792: 363	Curculionidae, Scolytinae [*Trypodendron signatum* (Fabricius, 1787)]
tiliae Panzer, 1793a: 14	Curculionidae, Scolytinae [*Ernoporus tiliae* (Panzer, 1793)]
volvulus Fabricius, 1792: 363 [ND]	Curculionidae, Scolytinae [current systematic position unknown]

Bostrichus Geoffroy, 1762

abietinus Fabricius, 1792: 367 [NN]	Curculionidae, Scolytinae [*Dendroctonus micans* (Kugelann, 1794)]
abietiperda Bechstein, 1818: 74	Curculionidae, Scolytinae [*Hylurgops palliatus* (Gyllenhal, 1813)]
abietis Ratzeburg, 1837: 163 [HN]	Curculionidae, Scolytinae [*Cryphalus asperatus* (Gyllenhal, 1813)]
abietis Ziegler in Sturm, 1826: 101 [NN HN]	Curculionidae, Scolytinae [*Cryphalus asperatus* (Gyllenhal, 1813)]
abietis Ziegler in Sturm, 1843: 230 [NN HN]	Curculionidae, Scolytinae [*Cryphalus asperatus* (Gyllenhal, 1813)]
albietis Ziegler in Dejean, 1821: 101 [NN]	Curculionidae, Scolytinae [*Cryphalus asperatus* (Gyllenhal, 1813)]
albietis Ziegler in Dejean, 1835: 307 [NN HN]	Curculionidae, Scolytinae [*Cryphalus asperatus* (Gyllenhal, 1813)]
acuminatus Gyllenhal, 1827: 620	Curculionidae, Scolytinae [*Ips acuminatus* (Gyllenhal, 1827)]
aesculi Kugelann, 1794: 525	Curculionidae, Scolytinae [*Hylesinus aesculi* (Kugellan, 1794)]
affaber Mannerheim, 1852: 359	Curculionidae, Scolytinae [*Dryocoetes affaber* (Mannerheim, 1852)]
alni Georg, 1856: 59	Curculionidae, Scolytinae [*Dryocoetes alni* (Georg, 1856)]
alni Mulsant et Rey, 1857a: 111 [HN]	Curculionidae, Scolytinae [*Xyleborus pfeilii* (Ratzeburg, 1837)]
alni Mulsant et Rey, 1857b: 111 [HN]	Curculionidae, Scolytinae [*Xyleborus pfeilii* (Ratzeburg, 1837)]
angustatus Herbst, 1784a: 111	Curculionidae, Scolytinae [*Hylastes angustatus* (Herbst, 1784)]

angustatus Sturm, 1826: 101 [NN HN] Curculionidae, Scolytinae [*Hylastes angustatus* (Herbst, 1784)]

angustatus Sturm, 1843: 230 [NN HN] Curculionidae, Scolytinae [*Hylastes angustatus* (Herbst, 1784)]

anomalus Schrank von Paula, 1795: 136 Staphylinidae, Pselaphinae [*Bryaxis bulbifer* (Reichernbach, 1816)]

aphodioides A. Villa et G. B. Villa, 1833: 36 Curculionidae, Scolytinae [*Crypturgus pusillus* (Gyllenhal, 1813)]

areccae Hornung, 1842: 117 Curculionidae, Scolytinae [*Hypothenemus areccae* (Hornung, 1842)]

asperatus Gyllenhal, 1813: 368 Curculionidae, Scolytinae [*Cryphalus asperatus* (Gyllenhal, 1813)]

asperatus Ratzeburg, 1837: 219 [HN] Curculionidae, Scolytinae [*Cryphalus saltuarius* (Weise, 1891)]

asperatus Sturm, 1826: 101 [NN HN] Curculionidae, Scolytinae [*Cryphalus saltuarius* (Weise, 1891)]

ater Fabricius, 1792: 368 Curculionidae, Scolytinae [*Pityogenes bidentatus* (Herbst, 1783)]

ater Paykull, 1800: 153 [HN] Curculionidae, Scolytinae [*Hylastes ater* (Paykull, 1800)]

autographus Ratzeburg, 1837: 160 Curculionidae, Scolytinae [*Dryocoetes autographus* (Ratzeburg, 1837)]

autographus Ullrich in Dejean, 1835: 307 [NN] Curculionidae, Scolytinae [*Dryocoetes autographus* (Ratzeburg, 1837)]

bicolor Herbst, 1784b: 116 Curculionidae, Scolytinae [*Taphrorychus bicolor* (Herbst, 1784)]

bicolor Chevrolat in Dejean, 1835: 307 [NN HN] Curculionidae, Scolytinae [*Taphrorychus bicolor* (Herbst, 1784)]

bidens Fabricius, 1792: 368 Curculionidae, Scolytinae [*Pityogenes bidentatus* (Herbst, 1783)]

bidentatus Herbst, 1783: 24 Curculionidae, Scolytinae [*Pityogenes bidentatus* (Herbst, 1783)]

binodulus Ratzeburg, 1837: 163 Curculionidae, Scolytinae [*Trypophloeus binodulus* (Ratzeburg, 1837)]

bispinus Duftschmid, 1825: 92 Curculionidae, Scolytinae [*Xylocleptes bispinus* (Duftschmid, 1825)]

bispinus Guyon, 1855: 4815 [HN] Curculionidae, Scolytinae [*Pityogenes bidentatus* (Herbst, 1783)]

boieldieui Perroud et Mountrouzier, 1864: 188	Curculionidae, Scolytinae [*Hypothenemus eruditus* (Westwood, 1836)]
brevis Panzer, 1796: 20	Curculionidae, Scolytinae [*Anisandrus dispar* (Fabricius, 1792)]
bulmerincqii Kolenati, 1846: 39	Curculionidae, Scolytinae [*Taphrorychus villifrons* (Dufour, 1843)]
calligraphus Duftschmid, 1825: 91	Curculionidae, Scolytinae [*Pityokteines curvidens* (Germar, 1824)]
carpophagus Hornung, 1842: 116	Curculionidae, Scolytinae [*Coccotrypes carpophagus* (Hornung, 1842)]
cavifrons Mannerheim, 1843: 297	Curculionidae, Scolytinae [*Trypodendron lineatum* (Olivier, 1790)]
cembrae Heer, 1836: 28	Curculionidae, Scolytinae [*Ips cembrae* (Heer, 1836)]
cinereus Herbst, 1784a: 116	Curculionidae, Scolytinae [*Crypturgus cinereus* (Herbst, 1784)]
compressicornis Fabricius, 1801: 358	Curculionidae, Scolytinae [*Cortylus compressicornis* (Fabricius, 1801)]
concinnus Mannerheim, 1852: 358	Curculionidae, Scolytinae [*Pseudoips concinnus* (Mannerheim, 1852)]
crenatus Fabricius, 1787: 37	Curculionidae, Scolytinae [*Hylesinus crenatus* (Fabricius, 1787)]
cristatus Fabricius, 1801: 389	Curculionidae, Scolytinae [*Dryocoetoides cristatus* (Fabricius, 1801)]
crudiae Panzer, 1791: 37	Curculionidae, Scolytinae [*Hypothenemus crudiae* (Panzer, 1791)]
cryptographus Ratzeburg, 1837: 160	Curculionidae, Scolytinae [*Xyleborus cryptographus* (Ratzeburg, 1837)]
cylindrus Fabricius, 1792: 364	Curculionidae, Platypodinae [*Platypus cylindrus* (Fabricius, 1792)]
dactyliperda Dejean, 1821: 101 [**NN HN**]	Curculionidae, Scolytinae [*Coccotrypes dactyliperda* (Fabricius, 1801)]
dactyliperda Fabricius, 1801: 387	Curculionidae, Scolytinae [*Coccotrypes dactyliperda* (Fabricius, 1801)]
dactyliperda Sturm, 1826: 101 [**NN HN**]	Curculionidae, Scolytinae [*Coccotrypes dactyliperda* (Fabricius, 1801)]
dentatus Sturm, 1826: 76 [**NN**]	Curculionidae, Scolytinae [*Ips pini* (Say, 1826)]

dentatus Sturm, 1843: 230 [**NN HN**]	Curculionidae, Scolytinae [*Ips pini* (Say, 1826)]
dispar Herbst, 1793: 113 [**HN**]	Curculionidae, Scolytinae [*Anisandrus dispar* (Fabricius, 1792)]
dryographus Ratzeburg, 1837: 167	Curculionidae, Scolytinae [*Xyleborus dryographus* (Ratzeburg, 1837)]
duplicatus C. R. Sahlberg, 1836a:144	Curculionidae, Scolytinae [*Ips duplicatus* (C. R. Sahlberg, 1836)]
duponti Montrouzier, 1861: 265	Curculionidae, Scolytinae [*Xyleborus perforans* (Wollaston, 1857)]
elongatus Fabricius, 1787: 36 [**HN**]	Zopheridae, Colydiinae [*Colydium elongatum*]
elongatus Herbst, 1784a: 117	Curculionidae, Scolytinae [*Hylurgus ligniperda* (Fabricius, 1787)]
euphorbiae Handschuch in Küster, 1845: 39	Curculionidae, Scolytinae [*Thamnurgus euphorbiae* (Handschuch in Küster, 1845)]
eurygraphus Ratzeburg, 1837: 168	Curculionidae, Scolytinae [*Xyleborus eurygraphus* (Ratzeburg, 1837)]
exesus Say, 1826: 255	Curculionidae, Scolytinae [*Ips calligraphus* (Germar, 1824)]
exiguus Walker, 1859: 260	Curculionidae, Scolytinae [*Xyleborinus exiguus* (Walker, 1859)]
exsculptus Ratzeburg, 1837: 162	Curculionidae, Scolytinae [*Pityophthorus exsculptus* (Ratzeburg, 1837)]
fagi Nördlinger, 1848: 242 [**ND**]	Curculionidae, Scolytinae [current systematic position unknown]
fasciatus Say, 1826: 255	Curculionidae, Scolytinae [*Monarthrum fasciatus* (Say, 1826)]
ferrugineus Boheman, 1858: 88 [**HN**]	Curculionidae, Scolytinae [*Xyleborus similis* (Ferrari, 1867)]
ferrugineus Fabricius, 1801: 388	Curculionidae, Scolytinae [*Xyleborus eurygraphus* (Ratzeburg, 1837)]
flavicornis Fabricius, 1777: 211	Curculionidae, Platypodinae [*Platypus cylindrus* (Fabricius, 1792)]
flavicornis Kugelann, 1792: 496 [**HN**]	Curculionidae, Platypodinae [*Platypus cylindrus* (Fabricius, 1792)]
flavipes Panzer, 1799: 9	Curculionidae, Scolytinae [*Hylurgus ligniperda* (Fabricius, 1787)]
flavipes Fabricius, 1801: 388 [**HN**]	Curculionidae, Scolytinae [*Taurodemus flavipes* (Fabricius, 1801)]

fraxini Panzer, 1799: 13 — Curculionidae, Scolytinae [*Hylesinus varius* (Fabricius, 1775)]

fraxini Bechstein in Bechstein et Sharfenberg, 1804: 107 [**HN**] — Curculionidae, Scolytinae [*Hylesinus varius* (Fabricius, 1775)]

frontalis Fabricius, 1801: 389 — Curculionidae, Scolytinae [*Pagiocerus frontalis* (Fabricius, 1801)]

fuscus Geoffroy in Fourcroy, 1785: 133 — Ptinidae [*Ptilinus fuscus*]

geminatus Sturm, 1843: 230 [**NN HN**] — Curculionidae, Scolytinae [*Ips acuminatus* (Gyllenhal, 1827)]

geminatus Zetterstedt, 1828: 345 — Curculionidae, Scolytinae [*Ips acuminatus* (Gyllenhal, 1827)]

granulatus Razteburg, 1837: 164 — Curculionidae, Scolytinae [*Trypophloeus granulatus* (Ratzeburg, 1837)]

histerinus Dufour, 1843: 91 — Curculionidae, Scolytinae [*Dryocoetes villosus villosus* (Fabricius, 1792)]

interruptus Mannerheim, 1852: 357 — Curculionidae, Scolytinae [*Ips tridens tridens* (Mannerheim, 1852)]

jalappae Letzner, 1849: 99 — Curculionidae, Scolytinae [*Scolytogenes jalappae* (Letzner, 1849)]

judeichii Kirsch, 1871: 388 — Curculionidae, Scolytinae [*Ips duplicatus* (C. R. Sahlberg, 1836)]

kaltenbachii Bach, 1850: 199 — Curculionidae, Scolytinae [*Thamnurgus kaltenbachi* (Bach, 1850)]

laricis Dejen, 1821: 100 [**NN HN**] — Curculionidae, Scolytinae [*Orthotomicus laricis* (Fabricius, 1792)]

laricis Fabricius, 1792: 365 — Curculionidae, Scolytinae [*Orthotomicus laricis* (Fabricius, 1792)]

laricis Sturm, 1826: 101 [**NN**] — Curculionidae, Scolytinae [*Orthotomicus laricis* (Fabricius, 1792)]

lichtensteinii Ratzeburg, 1837: 162 — Curculionidae, Scolytinae [*Pityophthorus lichtensteinii* (Ratzeburg, 1837)]

ligniperda Fabricius, 1787: 37 — Curculionidae, Scolytinae [*Hylurgus ligniperda* (Fabricius, 1787)]

limbatus Dejean, 1821: 101 [**NN HN**] — Curculionidae, Scolytinae [current systematic position unknown, probably *Trypodendron lineatum* (Olivier, 1795)]

limbatus Herbst, 1783: 24 — Curculionidae, Scolytinae [current systematic position unknown, probably *Trypodendron lineatum* (Olivier, 1795)]

lineatus Gyllenhal in Dejean, 1821: 101 [**NN HN**] — Curculionidae, Scolytinae [*Trypodendron lineatum* (Olivier, 1790)] *lineatus* Gyllenhal in Dejean, 1835: 307 [NN HN] Curculionidae, Scolytinae [*Trypodendron lineatum* (Olivier, 1790)]

lineatus Gyllenhall in Sturm, 1843: 230 [NN HN] Curculionidae, Scolytinae [*Trypodendron lineatum* (Olivier, 1790)]

lineatus Olivier, 1790b: 18 Curculionidae, Scolytinae [*Trypodendron lineatum* (Olivier, 1790)]

longicollis Gyllenhal, 1827: 621 Curculionidae, Scolytinae [*Orthotomicus longicollis* (Gyllenhal, 1827)]

mali Bechstein in Bechstein et Scharfenberg, 1805: 882 Curculionidae, Scolytinae [*Scolytus mali* (Bechstein in Bechstein et Scharfenberg, 1805)]

melanocephalus Fabricius, 1792: 368 Curculionidae, Scolytinae [*Hylesinus varius* (Fabricius, 1775)]

micans Kugelann, 1794: 523 Curculionidae, Scolytinae [*Dendroctonus micans* (Kugelann, 1794)]

micans Sturm, 1826: 101 [NN HN] Curculionidae, Scolytinae [*Dendroctonus micans* (Kugelann, 1794)]

micrographus Gyllenhal, 1813: 362 Curculionidae, Scolytinae [*Orthotomicus laricis* (Fabricius, 1792)]

micrographus Panzer in Sturm, 1843: 230 [NN HN] Curculionidae, Scolytinae [*Xyleborus dryographus* (Ratzeburg, 1837)]

minimus Fabricius, 1798: 158 Curculionidae, Scolytidae [*Carphoborus minimus* (Fabricius, 1798)]

minimus Fabricius in Dejean, 1821: 101 [NN] Curculionidae, Scolytidae [*Carphoborus minimus* (Fabricius, 1798)]

minutus Fabricius, 1777: 211 [ND] Curculionidae, Scolytinae [*Hylesinus varius* (Fabricius, 1775)]

minutus Fabricius, 1798: 211 [HN] Curculionidae, Scolytinae [*Hylesinus varius* (Fabricius, 1775)]

monocerus Gravenhorst, 1807: 193 [ND] Curculionidae, Scolytinae [current systematic position unknown]

monographus Fabricius, 1792: 365 Curculionidae, Scolytinae [*Xyleborus monographus* (Fabricius, 1792)]

nigritus Gyllenhal, 1827: 623 Curculionidae, Scolytinae [*Orthotomicus suturalis* (Gyllenhal, 1827)]

nitidulus Mannerheim, 1843: 298 Curculionidae, Scolytinae [*Pityophthorus nitidulus* (Mannerheim, 1843)]

octodentatus Gyllenhal in Dejean, 1821: 100 [NN HN] Curculionidae, Scolytinae [*Ips typographus* (Linnaeus, 1758)]

octodentatus Gyllenhal in Dejean, 1835: 307 [NN HN] Curculionidae, Scolytinae [*Ips typographus* (Linnaeus, 1758)] *octodentatus* Gyllenhal in Sturm, 1826: 101 [NN HN] Curculionidae, Scolytinae [*Ips typographus* (Linnaeus, 1758)]

octodentatus Gyllenhal in A. Villa et G. B. Villa, 1844: 463 [NN HN] Curculionidae, Scolytinae [*Ips typographus* (Linnaeus, 1758)]

octodentatus Paykull, 1800:146 Curculionidae, Scolytinae [*Ips typographus* (Linnaeus, 1758)]

oleae Fabricius, 1792: 366 Curculionidae, Scolytinae [*Phloeotribus scarabaeoides* (Bernard, 1788)]

oleiperda Fabricius, 1792: 366 Curculionidae, Scolytinae [*Hylesinus toranio* (D'Anthoine, 1788)]

orthographus Duftschmid, 1825: 91 Curculionidae, Scolytinae [*Pityokteines curvidens* (Germar, 1824)]

pallipes Sturm, 1826: 76 [NN] Curculionidae, Scolytinae [*Ips pini* (Say, 1826)]

palmicola Hornung, 1842: 116 Curculionidae, Scolytinae [*Coccotrypes dactyliperda* (Fabricius, 1801)]

parallelus Fabricius, 1801: 384 Curculionidae, Platypodinae [*Euplatypus parallelus* (Fabricius, 1801)]

pectinatus Laicharting, 1781: 68 Ptinidae [*Ptilinus pectinicornis* (Linnaeus, 1758)]

pfeilii Ratzeburg, 1837: 168 Curculionidae, Scolytinae [*Xyleborus pfeilii* (Ratzeburg, 1837)]

piceae Ratzeburg, 1837: 163 Curculionidae, Scolytinae [*Cryphalus piceae* (Ratzeburg, 1837)]

pinastri Bechstein, 1818: 74 [HN] Curculionidae, Scolytinae [*Ips sexdentatus* (Boerner, 1776)]

pinastri Bechstein in Bechstein et Sharfenberg, 1804: 93 Curculionidae, Scolytinae [*Ips sexdentatus* (Boerner, 1776)]

pinastri Sturm in Dejean, 1821: 100 [NN] Curculionidae, Scolytinae [*Ips sexdentatus* (Boerner, 1776)]

pinastri Sturm in Dejean, 1835: 307 [NN] Curculionidae, Scolytinae [*Ips sexdentatus* (Boerner, 1776)]

pini Say, 1826: 257 Curculionidae, Scolytinae [*Ips pini* (Say, 1826)]

pini Villa in Sturm, 1843: 230 [NN] Curculionidae, Scolytinae [*Ips pini* (Say, 1826)]

pityographus Ratzeburg, 1837: 162 Curculionidae, Scolytinae [*Pityophthorus pityographus pityographus* (Ratzeburg, 1837)] *plumeriae* Nördlinger, 1856: 74 — Curculionidae, Scolytinae [*Hypothenemus plumeriae* (Nördlinger, 1856)]

politus Say, 1826: 256 — Curculionidae, Scolytinae [*Xyloterinus politus* (Say, 1826)]

proboscidens Fabricius, 1801: 355 — Histeridae [*Coptotrphis proboscideus* (Fabricius, 1801)]

pubescens Fabricius, 1792: 368 — Curculionidae, Scolytinae [*Polygraphus poligraphus* (Linnaeus, 1758)]

pubipennis LeConte, 1860: 59 — Curculionidae, Scolytinae [*Pseudopityophthorus pubipennis* (LeConte, 1860)]

pumilio Gravenhorst, 1807: 193 [**ND**] — Curculionidae, Scolytinae [current systematic position unknown]

pusillus Gyllenhal, 1813: 371 — Curculionidae, Scolytinae [*Crypturgus pusillus* (Gyllenhal, 1813)]

pusillus Gyllenhal in Dejean, 1821: 101 [**NN HN**] — Curculionidae, Scolytinae [*Crypturgus pusillus* (Gyllenhal, 1813)]

pusillus Gyllenhal in Dejean, 1835: 307 [**NN HN**] — Curculionidae, Scolytinae [*Crypturgus pusillus* (Gyllenhal, 1813)]

pusillus Gyllenhal in Sturm, 1826: 101 [**NN HN**] — Curculionidae, Scolytinae [*Crypturgus pusillus* (Gyllenhal, 1813)]

pusillus Stéven in Dejean, 1835: 307 [**NN HN**] — Curculionidae, Scolytinae [*Crypturgus pusillus* (Gyllenhal, 1813)]

pygmaeus Fabricius, 1787: 37 — Curculionidae, Scolytinae [*Scolytus pygmaeus* (Fabricius, 1787)]

quadridens T. Hartig in G. L. Hartig et T. Hartig, 1834: 109 — Curculionidae, Scolytinae [*Pityogenes quadridens* (T. Hartig in G. L. Hartig et T. Hartig, 1834)]

quinquelineatus Adams, 1817: 312 — Curculionidae, Scolytinae [*Trypodendron signatum* (Fabricius, 1787)]

ratzeburgii Kolenati, 1846: 39 — Curculionidae, Scolytinae [*Anisandrus dispar* (Fabricius, 1792)]

ruficollis Fabricius, 1801: 388 — Curculionidae, Scolytinae [*Tricolus ruficollis* (Fabricius, 1801)]

rugulosus P. W. J. Müller, 1818: 247 — Curculionidae, Scolytinae [*Scolytus rugulosus* (P. W. J. Müller, 1818)]

saxesenii Ratzeburg, 1837: 167 — Curculionidae, Scolytinae [*Xyleborinus saxesenii* (Ratzeburg, 1837)]

scolytus Fabricius, 1775: 59	Curculionidae, Scolytinae [*Scolytus scolytus* (Fabricius, 1775)]
semicastaneus Mannerheim, 1852: 328	Curculionidae, Scolytinae [*Dryocoetes autographus* (Ratzeburg, 1837)]
septentrionis Mannerheim, 1843: 298	Curculionidae, Scolytinae [*Dryocoetes autographus* (Ratzeburg, 1837)]
serratus Fabricius, 1801: 386 [HN]	Curculionidae, Scolytinae [*Taurodemus varians* (Fabricius, 1801)]
serratus Panzer, 1795: 288	Curculionidae, Scolytinae [*Ernoporicus fagi* (Fabricius, 1798)]
sidneyanus Nördlinger, 1856: 75	Curculionidae, Scolytinae [*Cryphalus sidneyanus* (Nördlinger, 1856)]
signatus Dejean 1821: 101 [NN]	Curculionidae, Scolytinae [*Trypodendron signatum* (Fabricius, 1787)]
signatus Fabricius in Dejean 1835: 307 [NN HN]	Curculionidae, Scolytinae [*Trypodendron signatum* (Fabricius, 1787)]
signatus Fabricius in Sturm, 1843: 230 [NN HN]	Curculionidae, Scolytinae [*Trypodendron signatum* (Fabricius, 1787)]
stenographus Creutzer in Dejean, 1821: 100 [NN]	Curculionidae, Scolytinae [*Ips sexdentatus* (Boerner, 1876)]
stenographus Creutzer in Dejean, 1835: 307 [NN HN]	Curculionidae, Scolytinae [*Ips sexdentatus* (Boerner, 1876)]
stenographus Duftschmid, 1825: 88	Curculionidae, Scolytinae [*Ips sexdentatus* (Boerner, 1876)]
suturalis Gyllenhal, 1827: 622	Curculionidae, Scolytinae [*Orthotomicus suturalis* (Gyllenhal, 1827)]
tachygraphus C. R. Sahlberg, 1836: 152	Curculionidae, Scolytinae [*Anisandrus dispar* (Fabricius, 1792)]
terminalis Mannerheim, 1843: 298	Ciidae (Wood 1969); according to Borowski & Węgrzyxnowicz, 2007 as *Cryphalus terminalis*, but Wood & Bright (1992) and Alonso-Zarazaga (2023) this species in subfamily Scolytinae not mentioned
testaceus Fabricius, 1787: 37 [*Bostricilus* sic! – LC]	Curculionidae, Scolytinae [*Tomicus piniperda* (Linnaeus, 1758)]

testaceus Walker, 1859: 260 [HN] Curculionidae, Scolytinae [*Xyleborus perforans* (Wollaston, 1857)]

thoracicus Panzer, 1796: 18 Curculionidae, Scolytinae [*Anisandrus dispar* (Fabricius, 1792)]

thoracicus Fabricius, 1801: 385 [HN] Histeriade [*Trypanaeus thoracicus* (Fabricius, 1801)]

trepanatus Nördlinger, 1848: 239 Curculionidae, Scolytinae [*Pityogenes trepanatus* (Nördlinger, 1848)]

tridens Mannerheim, 1852: 357 Curculionidae, Scolytinae [*Ips tridens tridens* (Mannerheim, 1852)]

trifolii P. W. Müller, 1807: 56 Curculionidae, Scolytinae [*Hylastinus obscurus* (Marsham, 1802)]

tuberculosus Herbst, 1784a: 113 Curculionidae, Scolytinae [*Xyleborus monographus* (Fabricius, 1792)]

typographus Dejean, 1821: 100 [NN HN] Curculionidae, Scolytinae [*Ips typographus* (Linnaeus, 1758)]

typographus Fabricius in Sturm, 1843: 230 [NN HN] Curculionidae, Scolytinae [*Ips typographus* (Linnaeus, 1758)]

typographus Fabricius in A. Villa et G. B. Villa, 1844: 463 [NN HN] Curculionidae, Scolytinae [*Ips typographus* (Linnaeus, 1758)]

typographus Sturm, 1826: 110 [NN HN] Curculionidae, Scolytinae [*Ips typographus* (Linnaeus, 1758)]

typographus Ziegler in Dejean, 1835: 307 [NN] Curculionidae, Scolytinae [*Ips typographus* (Linnaeus, 1758)]

unidentatus Fabricius, 1801: 386 Curculionidae, Scolytinae [*Taurodemus unidentatus* (Fabricius, 1801)]

varians Fabricius, 1801: 386 Curculionidae, Scolytinae [*Taurodemus varians* (Fabricius, 1801)]

varius Fabricius, 1775: 60 Curculionidae, Scolytinae [*Hylesinus varius* (Fabricius, 1775)]

victoris Mulsant et Rey, 1853: 91 Curculionidae, Scolytinae [*Dryocoetes autographus* (Ratzeburg, 1837)]

villifrons Dufour, 1843: 91 Curculionidae, Scolytinae [*Taphrorychus villifrons* (Dufour, 1843)]

villosus Fabricius, 1792: 367 Curculionidae, Scolytinae [*Dryocoetes villosus villosus* (Fabricius, 1792)]

villosus Herbst, 1793: 121 [HN] Curculionidae, Scolytinae [*Dryocoetes autographus* (Ratzeburg, 1837)]

villosus Ratzeburg, 1837: 160 [HN] Curculionidae, Scolytinae [*Xyleborus cryptographus* (Ratzeburg, 1837)]

vittatus Fabricius, 1787: 368	Curculionidae, Scolytinae [*Pteleobius vittatus* (Fabricius, 1787)]
volvulus Fabricius, 1775: 454	Curculionidae, Scolytinae [*Xyleborus volvulus* Fabricius, 1775]
waringii Curtis, 1840: 279	Curculionidae, Scolytinae [*Trypodendron signatum* (Fabricius, 1787)]
xanthographus Say, 1826: 256	Curculionidae, Scolytinae [*Xyleborus*? – current systematic position unknown]
xylographus C. R. Sahlberg, 1836b: 148 [HN]	Curculionidae, Scolytinae [*Pityogenes chalcographus* (Linnaeus, 1761)]
xylographus Say, 1826: 256	Curculionidae, Scolitinae [*Xyleborus xylographus* (Say, 1826)]

Bostrychus Agassiz, 1846 [LC] (= *Bostrichus* Geoffroy, 1762)

autographus Knoch in Gistel, 1856 368 [NN]	Curculionidae, Scolytinae [*Dryocoetes autographus* (Ratzeburg, 1837)]
carinatus Waltl, 1839: 223 [ND]	Curculionidae, Scolytinae [current systematic position unknown]
delphinii Rosenhauer, 1856: 302	Curculionidae, Scolytinae [*Thamnurgus delphinii* (Rosenhauer, 1856)]
eudromius Gistel, 1857: 33 [NN]	Curculionidae, Scolytinae [*Trypodendron signatum* (Fabricius, 1787)]
maculatus Lacepède, 1802: 143	Animalia, Chordata, Channidae
parenchymatis Gistel, 1857: 41 [NN]	Curculionidae, Scolytinae [*Trypodendron signatum* (Fabricius, 1787)]
pilosus Dejean, 1821: 101 [NN]	Curculionidae, Scolytinae [*Trypodendron signatum* (Fabricius, 1787)]
pilosus Dejean, 1835: 307 [NN]	Curculionidae, Scolytinae [*Trypodendron signatum* (Fabricius, 1787)]
pilosus Gistel, 1831: 306 [ND]	Curculionidae, Scolytinae [current systematic position unknown]
pilosus Gmelin, 1790: 1603 [ND]	Curculionidae, Scolytinae [current systematic position unknown]
pinetarius Gistel, 1857: 517 [NN]	Curculionidae, Scolytinae [*Trypodendron signatum* (Fabricius, 1787)]

sinensis Lacepède, 1802: 145	Animalia, Chordata, Channidae
westerhauseri Gistel, 1857: 73	Curculionidae, Scolytinae [*Trypodendron signatum* (Fabricius, 1787)]

Lyctus Fabricius, 1792

abbreviatus Panzer, 1794a: 21	Nitidulidae [*Carpophilus sexpustulatus* (Fabricius, 1791)]
bipustulatus Fabricius, 1792: 503	Monotomidae [*Rhizophagus bipustulatus* (Fabricius, 1792)]
contractus Fabricius, 1792: 505	Bothrideridae [*Bothrideres bipunctatus* (Gmelin, 1790)]
contractus Strum, 1826: 166 [NN HN]	Bothrideridae [*Bothrideres bipunctatus* (Gmelin, 1790)]
cylindricus Creutzer in Panzer, 1796: 18	Bothrideridae [*Oxylaemus cylindricus* (Creutzer in Panzer, 1796)]
depressiusculus White et Butler, 1846: 18	Zopheridae [*Pycnomerus depressiusculus* (White et Butler, 1846)]
depressus Fabricius, 1792: 503	Monotomidae [*Rhizophagus depressus* (Fabricius, 1792)]
dermestoides Panzer, 1793b: 15	Nitidulidae [*Pityophagus ferrugineus* (Linnaeus, 1761)]
dispar Paykull, 1800: 328	Monotomidae [*Rhizophagus dispar* (Paykull, 1800)]
duodecimstriatus P. W. J. Müller, 1821: 190	Bothrideridae [*Anommatus duodecimstriatus* (P. W. J. Müller, 1821)]
fasciculosus Gyllenhal, 1827: 632	Zopheridae [*Xylolaemus fasciculosus* (Gyllenhal, 1827)]
ferrugineus Paykull, 1800: 326	Monotomidae [*Rhizophagus ferrugineus* (Linnaeus, 1758)]
haematodes Fabricius, 1801: 62	Zopheridae [*Pycnomerus haematodes* (Fabricius, 1801)]
haematodes Say, 1826: 262 [HN]	Zopheridae [*Pycnomerus haematodes* (Fabricius, 1801)]
histeroides Fabricius, 1792: 295	Cerylonidae [*Cerylon histeroides* (Fabricius, 1792)]
nitidulus Fabricius, 1798: 177	Monotomidae [*Rhizophagus nitidulus* (Fabricius, 1798)]
nitidus Fabricius, 1792: 505	Bothrideridae [*Teredus cylindricus* (Olivier, 1790)]

obscurus Fabricius, 1801: 562	Curculionidae, Scolytinae [*Hypotenemus obscurus* (Fabricius, 1801)]
parvulus Paykull, 1800: 329	Monotomidae [*Rhizophagus fenestralis* (Linnaeus, 1758)]
reflexus Say, 1826: 262	Zopheridae [*Pycnomerus reflexus* (Say, 1826)]
rufipennis Montrouzier, 1861: 268	Zopheridae [*Pycnomerus rufipennis* (Montrouzier, 1861)]

Xylotrogus Stephens, 1830 (= _Lyctus_ Stephens, Fabricius, 1792)

brevicornis Melsheimer, 1846: 112	Zopheridae [*Pycnomerus haematodes* (Fabricius, 1801)]

List of Bostrichidae incertae sedis (probably not Bostrichidae but other Families), their current systematic position unknown (probably *nomina nuda* or *nomina dubia*)

Apate Fabricius, 1775

brunipennis Fabricius, 1801: 383 [ND]	
gigantea Lichtenstein, 1796: 25 [ND]	probably Lymexylonidae or Zopheridae (according to author of description)
glabrata Fabricius, 1801: 383 [ND]	
nigricans Panzer, 1797: 95 [ND]	
pectinata Grimmer, 1841. 44 [ND]	
perniciosa Gistel, 1857: 5 [NN]	
punctata Lichtenstein, 1796: 26 [ND]	
viridis Panzer, 1797: 95 [ND]	

Bostrichus Geoffroy, 1762

alpinus Grimmer, 1841: 44 [ND]	
bicornis Thunberg, 1798: 114 [ND]	probably as *Apate* – footnote according to author of description
bifasciatus Gmelin, 1790: 1603 [ND]	
bipustulatus Fabricius, 1801: 385 [ND]	
boleti Kugelann, 1792: 497 [ND]	
bruchoides Rossi, 1790: 38 [ND]	
calcographus Olivier in A. Villa et G. B. Villa, 1844: 463 [NN]	
dentatus Fabricius, 1792: 503 [ND]	
diodon Schüppel in A. Villa et G. B. Villa, 1844: 463 [NN]	

elongatus Gyllenhal in A. Villa et G. B. Villa, 1844: 463 [NN]
furcatus Bosc, 1792: 259 (*Bostricus* sic!) [ND]
micrographus Dejean, 1821: 101 [NN HN]
micrographus Fabricius in Dejean, 1836: 307 [NN HN]
micrographus Sturm, 1826: 102 [NN HN]
micrographus Ziegler in Dejean, 1821: 101 [NN]
micrographus Ziegler in Dejean, 1835: 307 [NN HN]
ortographus Creutzer in A. Villa et G. B. Villa, 1844: 463 [NN]
ruficornis Schrank von Paula, 1798: 437 [ND]
sulcicollis R. A. Philippi & H. E. Philippi, 1864: 375 [ND]

Bostrychus Agassiz, 1846
bidens Gistel, 1856: 368 [NN]
chalcographus Gistel, 1856: 368 [NN]
curvidens Georg in Gistel, 1856: 368 [NN]
laricis Gistel, 1856: 368 [NN]
micrographus Panzer in Gistel, 368 [NN]
suturalis Ratzeburg in Gistel, 1856: 368 [NN]

Lyctus Fabricius, 1792
aeneus Richter, 1820: 9 [ND]
brunneus Fabricius, 1792: 503 [ND]
dentatus Fabricius, 1792: 503 [ND]
testaceus Drapiez, 1819: 294 [ND]

References

[1]Primary sources with description of new taxa.
[2]Primary sources with description of new taxa – species probably not belonging to the family Bostrichidae.
[3]Primary sources with description of new taxa – species not belonging to family Bostrichidae (despite being described in bostrichid genera).
Without any "index" – sources important for synonymy, new combination or distribution.

[3]Adams M. 1817: Descriptio insectorum novorum Imperii Rossici, imprimis Caucasi et Siberiae. *Memoires de la Sociéte Impériale des Naturalistes Moscou* 5: 278–314.

[1]Agassiz L. 1846a: *Nomenclatoris Zoologici index universalis, continens nomina systematica classium, ordinum, familiarum et generum animalium omnium, tam viventiuum quam fossilium, secundum ordinem alphabeticum unicum disposita, adjectis homonymiis plantarum, nec non variis adnotationibus et emendationibus*. Soloduri, 1–49.

Agassiz L. 1846b: *Nomina systematica generum Coleopterorum. Tom viventium quam fossilium. Secundum ordinem alphabeticum disposita adjectis auctoribus, libris, in quibus reperiuntur, anno editionis, etymologia et familias ad quas pertinent, in singulis classibus*. Soloduri, xi + [2] + 170. *In: Agassiz L (1842–1846). Nomenclator Zoologicus continens nomina systematica generum animaliutm tam viventium quam fossilium*, (Paginated separately according to individual taxonomy groups, 1556 pp.).

[1]Allard E. 1869: Description de quelques Coléoptères nouveaux et notes diverses. *L'Abeille Mémoires d'Entomologie* 5 [1868–1869]: 465–478.

Alonso-Zarazaga M. A. 2023: *Cooperative Catalogue of Palaearctic Coleoptera. Curculionoidae. 2nd edition*. Monographias electrónicas SEA, Sociedad Entomológica Aragonesa S.E.A., Zaragosa, Volume 14, 780 pp.

Al-Safar H. H. & Augul R. S. 2022: Survey and revision of storage insects from several localities of Iraq. GSC Biological and Pharmaceutical Sciences. Doi.org/10.30574/gsbps.2022.20.3.0351.

An Y. 2012: Wai lai senlin youhai shengwu qianyi [*Quarantine inspection of pests from foreign forests*] (*Chinese Edition*). Peking: Science Press, 695 pp. (in Chinese).

[1]Ancey C. M. F. 1879: Description d'une nouvelle espèce de Coléoptère du genre Sinoxylon. *Le Naturaliste, Journal des Échanges et des Nouvelles* 1: 139–140.

[1]Ancey C. M. F. 1881: Descriptions de Coléoptères nouveaux. *Le Naturaliste, Journal des Échanges et des Nouvelles* 3: 509.

[1]Anonymous 1939: Insecta. *Zoological Record* [1938] 75(12): 1–427.

Arrow G. J. 1904: Notes on two species of Coleoptera introduced into Europe. *The Entomologist's Monthly Magazine* (Series 2), 15(40): 35–36.

Bach M. 1850: Weiteres über Bostrichus Kaltenbachii. *Entomologische Zeitung* [Stettin]: 11(1): 18–19.

[1)]Bach M. 1852: *Käferfauna für Nord- und Mitteldeutschland, mit besonderer Rücksicht auf die preussischen Rheinlande. Zweiter Band. 3.* Coblenz: J. Hölscher, 6 + 148 pp.

Bahillo de la Puebla P., López-Colón J. I. & Baena M. 2007: Los Bostrichidae Latreille, 1802 de la fauna íbero-balear (Coleoptera). *Heteropterus Revista de Entomología* 7(2): 147–227.

[1)]Basilewsky P. 1952: Les Bostrychides du Congo Belge (Coleoptera, Cucujoidea). *Revue de Zoologique et de Botanique Africaines* 46: 81–150.

[1)]Basilewsky P. 1954: Description d'un Coléoptère Bostrichidae nouveau de l'Afrique Centrale. *Revue de Zoologie et de Botanique Africaines* 49: 77–80.

[1)]Basilewsky P. 1955: Contributions à l'étude de la faune entomologique du Ruanda-Urundi (Mission P. Basilewsky 1953) XIII. Coleoptera Lyctidae et Bostrychidae. *Annales du Musée Royal du Congo Belge, Sciences Zoologiques* 36(1): 135–144.

Baudi di Selve F. 1873: Catalogo dei Dascillidi, Malacodermi e Teredili della Fauna europea e circummediterranea appartenenti alle collezioni del Museo Civico de Genova. *Annali della Museo di Storia Naturale di Genova* 4: 226–268.

[1)]Baudi di Selve F. 1874: Coleopterorum messis in insula Cypro et Asia minore ab Eugenio Truqui congregatae recensitio: de Europaeis notis quibusdam additis. Pars quinta. *Berliner Entomologische Zeitschrift* 17: 317–338.

Beaver R. A., Sittichaya W. & Liu L. Y. 2011: A review of the powder-post beetles of Thailand (Coleoptera: Bostrichidae). *Tropical Natural History* 11: 135–158.

[3)]Bechstein J. M. 1818: *Forstinsektologie oder Naturgeschichte der für den Wald schädlichen und nützlichen Insecten, nebst Einleitung in die Insektenkunde überhaupt für angehende und ausübende Forstmänner und Cameralisten.* Gotha: Hennings, xii + 551 + 8 pp. + 4 pl.

[3)]Bechstein J. M. & Scharfenberg G. L. 1804: *Vollständige Naturgeschichte der schädlichen Forstinsekten. Ein Handbuch für Forstmänner, Cameralisten und Oekonomen. Erster Theil.* Leipzig: C. F. Knoch Richter. VIII + 292 pp. + 1 [unn. Verzeichniss der Kupfertafeln] + 3 pl.

[3)]Bechstein J. M. & Scharfenberg G. L. 1805: *Vollständige Naturgeschichte der schädlichen Forstinsekten. Ein Handbuch für Forstmänner, Cameralisten und Desonomen. Dritter Theil.* Leipzig: C. F. Knoch Richter, 605–1046 + [4] pp. + pls. 10–13.

[1)]Beeson C. F. C. & Bhatia B. M. 1937: On the biology of the Bostrichidae (Ceóleopt.). *Indian Forest Records (New Series) Entomology* 2: 223–323.

[1)]Belkin J. N. 1940: Notes on North American Bostrichidae (Coleoptera). *Entomological News* 51: 192–193.

[1)]Billberg G. J. 1820a: *Enumeratio Insectorum in Museo Gust. Joh. Billberg.* Stockholm: Gadelianis, [2] + 138 pp.

[1])Billberg G. J. 1820b: Novae insectorum species, descriptae. *Mémoires de l'Académie Impériale de Sciences de St-Petersbourgh* 7 [1815–1816]: 381–395.

Binda F. & Joly L. J. 1991: Los Bostrichidae (Coleoptera) de Venezuela. *Boletín de Entomología Venezolana N.S.* 6(2): 83–133.

[1])Blackburn T. 1888: Further Notes on Australian Coleoptera, with Descriptions of New Species. *Transaction and Proceedings and Report of the Royal Society of South Australia* 10 [1886–1887]: 177–287.

[1])Blackburn T. 1889: Further Notes on Australian Coleoptera with descriptions of New Genera and Species. *The Proceedings of the Linnean Society of New South Wales* (2) 3 [1888]: 1387–1506.

[1])Blackburn T. 1890: Notes on Australian Coleoptera with Descriptions of New Species. V. *Proceedings of the Linnaean Society of New South Wales* (2) 4 [1889]: 1247–1276.

[1])Blackburn T. 1892: Further Notes on Australian Coleoptera, with Descriptions of New Genera and Species. XI. *Transactions and Proceedings of the Royal Society of South Australia* 15(1) [1891–1892]: 20–73.

[1])Blackburn T. 1893: Further Notes of Descriptions of New Genera and Species. XIII. *Transactions and Proceedings of the Royal Society of South Australia* 17(1): 130–140.

[1])Blackburn T. 1897: Further Notes of Descriptions of New Genera and Species. XXII. *Transactions and Proceedings of the Royal Society of South Australia* 21(2) [1896–1897]: 88–98.

[1])Blackburn T. & Sharp D. 1885: Memoirs on the Coleoptera of the Hawaiian Islands. *The Scientific Transaction of the Royal Dublin Society* (2) 3: 119–289 + 2 pl.

Blackwelder R. E. 1945: Checklist of the Coleopterous Insects of Mexico, Central America, the West Indies, and South America. Part 3. *Smithsonian Institute United States National Museum. Bulletin* 185: 343–550.

Blair K. G. 1928a: Heteromera, Bostrychoidea, Malacodermata and Buprestidae. Pp. 67–109. In: Insects of Samoa and other Samoan terrestrial Arthropoda. Part. IV. Coleoptera, Fasc. 2. *British Museum. London Natural History* 4(2): 67–174.

Blair K. G. 1928b: Coleoptera (Heteromera, Teredilia, Malacodermata, and Bruchidae) from the Galapagos Islands, collected on the 'St. George' Expedition, 1924. *The Annals and Magazine of Natural History* (10) 1: 671–680.

Blair K. G. 1935: Further new species and other records of Marquesan Coleoptera. In: Marquesan Insects – II. *Bulletin Bernice P. Bishop Museum Honolulu* 114: 289–297.

[1])Blanchard C. É. 1851: Coleópteros: Tetramerés, Trimeros, Dimeros [pp. 285–564, pls. 22–32]. In: Gay C. (ed): *Historia física y política de Chile Zoología Tomo quinto*. Paris: Privately published, 563 pp.

[1])Blanchard C. É. & Brullé A. 1843: *Insectes l'Amérique Méridionale recueillis par Alcide d'Orbigny, et décrits par Émile Blanchard et August Brullé*. Pp. 57–222. In: Blanchard C. É.: *Voyage dans l'Amérique Méridionale (le Brésil, le République Orientale de l'Urugay, la République Argentine, la Patagoniae, la République du Chili, la*

République de Bolivia, la République du Pérou), exécuté pedants les anées 1826, 1827, 1828, 1829, 1830, 1831, 1832 et 1833 par Alcide d'Orbigny. Tome sixième. 1e Partie: Insectes. (Excl. Carabiques, Hydrocanthares, Palpicornes). Paris, Strasbourgh, [4] + 222 pp. + 32 pls.

3)Boheman C. H. 1858: Coleoptera. Species noveas descripsit. In: *Kongliga Svenska Fregatten Eugenies Resa Omkring Jorden under befäl af C. A. Virgin aren 1851–1853. Vetenskapliga Iakttaagelser. II. Zoologi. 1. Insecta*. Uppsala & Stockholm: Almquist & Wiksells, pp. 1–112.

1)Boisduval J. B. A. D. de 1835: *Voyage de Découvertes l'Astrolabe exécuté par order du Roi, pendant les Anées 1826–1827–1829–1829, sous le commandement de M. J. Dumont d'Urville. Faune entomologique de l'Océan Pacifique, avec l'illustration des insectes nouveaux recueillis pendant le voyage. Deuxième partie. Coléoptères et autres ordres*. Paris: J. Tastu, viii + 716 pp.

Boriani M., Taddei A., Bazzoli M. & Michelotti S. 2019: Detection and molecular analysis of three exotic auger beetles of the genus Sinoxylon Duftschmidt, 1825 (Coleoptera: Bostrichidae) intercepted in Italy. *Polish Journal of Entomology* 88(1): 1–13.

Borowski J. 2007: Bostrichidae. Pp. 320–328. In: Löbl I. & Smetana A. (eds.): *Catalogue of Palaearctic Coleoptera. Elateroidea – Derodontoidea – Bostrichoidea – Lymexyloidea – Cleroidea – Cucujoidea. Volume 4*. Stenstrup: Apollo Books, 935 pp.

1)Borowski J. 2018: Materials to the knowledge of Bostrichidae (Coleoptera) of The Republic of Gambia. *World Scientific News* 106: 1–11.

1)Borowski J. 2020: World inventory of beetles of the family Bostrichidae (Coleoptera). Part 1. Check List from 1758 to 2007. *World Scientific News* 140(1): 1–49.

Borowski J. 2022: Taxonomic changes in Bostrichidae (Insecta: Coleoptera). *Zootaxa* 5154(5): 590–594.

1)Borowski J., Lasoń A. & Sławski M. 2021: Remarks on Xylomedes Lesne, 1902 with description of a new species from Iran (Coleoptera: Bostrichidae: Apatinae). *Zootaxa* 4941(2): 291–300.

Borowski J. & Singh S. 2017: Bostrichidae and Ptinidae: Ptininae (Insecta: Coleoptera) type collection at National Forest Insect Collection, Forest Research Institute, Dehradun (India). *World Scientific News* 66: 193–224.

1)Borowski J. & Sławski M. 2017: Bostrichidae (Coleoptera) of Socotra with descriptions of two new subspecies. *Acta Entomologica Musei Nationalis Pragae* (supplementum) 57: 101–111.

Borowski J. & Węgrzynowicz P. 1999: Bostrichidae and Ptinidae (Bostrichoidea, Coleoptera). Results of Zoological Expeditions of Museum and Institute of Zoology PAS in Vietnam (1996–1999). *Bulletin of the Museum and Institute of Zoology PAS, supplement to Annales Zoologici Warszawa* 1999(2): 125–126.

1)Borowski J. & Węgrzynowicz P. 2007a: *World Catalogue of Bostrichidae (Coleoptera)*. Mantis, Olsztyn, 247 pp. + 8 pls.

Borowski J. & Węgrzynowicz P. 2007b: New synonym of Apoleon edax Gorham, 1885 (Coleoptera: Bostrichidae). *Genus* 18: 367–369.

Borowski J. & Węgrzynowicz P. 2008: Calophagus pekinensis Lesne, 1902, a Bostrichid Beetle new to the Japanese Fauna. *Elytra* 36(2): 285–286.

Borowski J. & Węgrzynowicz P. 2009a: Megabostrichus imadatei Chûjô, 1964 (Coleoptera, Bostrichidae), a new synonym of Phonapate fimbriata Lesne, 1909. *Elytra* 37(2): 287–288.

Borowski J. & Węgrzynowicz P. 2009b: Stephanopachys sachalinensis (Matsumura, 1911) (Coleoptera, Bostrichidae), a new synonym of Stephanopachys substriatus (Paykull, 1800). *Elytra* 37(2): 289–290.

Borowski J. & Węgrzynowicz P. 2009c: Apate Fabricius, 1775 (Bostrichidae: Coleoptera): A protected name. *Annales Zoologici* 59(2): 189–191.

Borowski J. & Węgrzynowicz P. 2009d: Calophagus pekinensis Lesne, 1902, a Bostrichid beetle new to the Japanese fauna. *Elytra* 37(2): 285–286.

Borowski J. & Węgrzynowicz P. 2009e: Xylopsocus intermedius Damoiseau, 1993 (Coleoptera, Bostrichidae), a New Beetle in the Taiwanese and Palearctic Fauna. *Elytra* 37: 103–104.

Borowski J. & Węgrzynowicz P. 2009f: Xylopsocus galloisi Lesne, 1937 (Coleoptera, Bostrichidae), a New Beetle in Chinese Fauna. *Elytra* [Tokyo] 37: 295–296.

Borowski J. & Węgrzynowicz P. 2010: Acantholyctus semiermis (Lesne, 1914), a new synonym of Acantholyctus cornifrons (Lesne, 1898) (Coleoptera: Bostrichidae). *Genus* 21: 501–504.

Borowski J. & Węgrzynowicz P. 2011a: New Bostrichid beetles to the Taiwanese fauna, with a key for identification of the genus Dinoderus of Taiwan. *Elytra, New Series* 1(1): 93–96.

[1)]Borowski J. & Węgrzynowicz P. 2011b: Orientoderus a new subgenus of Prostephanus Lesne, 1897, with description of a new species from Thailand and Laos (Coleoptera, Bostrichidae). *Elytra, New Series* 1(2): 255–261.

Borowski J. & Węgrzynowicz P. 2012: *The powderpost beetles of the world (Coleoptera: Bostrichidae): Keys for the identification of species. Volume 1*. Mantis: Olsztyn, 461 pp. + 16 pls.

[1)]Borowski J. & Węgrzynowicz P. 2013: A new species of the genus Dinoderus Stephens from Taiwan (Coleoptera: Bostrichidae, Dinoderinae). *Genus* 24(1): 33–38.

Borowski J. & Węgrzynowicz P. 2019: Nomenclatural changes in the genus Xylothrips Lesne, 1901 (Coleoptera, Bostrichidae, Xyloperthini). *World News of Natural Sciences* 24: 79–88.

Borowski T. 2020: World inventory of beetles of the family *Bostrichidae* (Coleoptera). Part 1. Check List from 1758 to 2012. *World News of Natural Science* 28: 155–170.

Borowski T. 2021: World inventory of beetles of the family Bostrichidae (Coleoptera). Part 2. Check List from 1758 to 2007. *World News of Natural Science* 36: 9–41.

[2])Bosc [d'Antic] L. A. G. 1792: Bostricus furcatus. *Journal d'Histoire Naturelle* 2: 259–260.

Bouchard P., Bousquet Y., Davies A. E., Alonso-Zarazaga M. A., Lawrence J. F. Lyal, C. H. C., Newton A. F., Reid C. A. M., Schmitt M., Ślipińskin A. & Smith A. B. T. 2011: Family-group names in Coleoptera (Insecta). *ZooKeys* 88: 1–972.

Bousquet Y. 1990: *Beetles associated with stored products in Canada: An identification guide.* Ottawa: Canadian Government Publishing Centre, iv + 215 pp.

Bousquet Y. 1991: Superfamily Bostrichoidea. In: Bousquet Y.: *Checklist of beetles of Canada and Alaska.* Ottawa: Biosystematics Research Centre, vi + 430 pp.

Bousquet Y. 2016: Litteratura Coleopterologica (1758–1900): a guide to selected book related to the taxonomy of Coleoptera with publication dates and notes. *ZooKeys* 583: 1–776.

Bousquet Y., Bouchard P., Davies A. E., Sikes D. S. 2013: *Checklist of beetles (Coleoptera) of Canada and Alaska. Second edition.* Sofia-Moscow: Pensoft, 402 pp.

[1])Bradley J. C. 1930: *A manual of the genera of beetles of America north of Mexico. Keys for the determination of the families, subfamilies, tribes, and genera of Coleoptera with a systematic list of the genera and higher groups.* Ithaca: Daw, Illston & Co., x + 360 pp.

[1])Brancsik K. 1893: Beiträge zur Kenntnis Nossibés und dessen Fauna nach Sendungen und Mittheilungen des Herrn P. Frey. *Jahresberichte Naturwisenschaftichen Vereinnes Trencsin Comitates* [A Trencsén Vármegyei Témészttudományi Egylet] 15–16: 202–258.

[1])Broun T. 1880: *Manual of the New Zealand Coleoptera.* Wellington: J. Hughes, xx + 652 pp.

[1])Bruch C. 1915: Catálogo sistemático de los Coleópteros de la República Argentina. Pars VI. *Revista del Museo de la Plata* 19: 235–302.

Brustel H. & Aberlenc H.-P. 2014: Les Bostrichidae Latreille, 1802 de la faune de France: espèces autochtones, interceptées, introduites ou susceptibles de l'être (Coleoptera). *Revuede l'Association Roussillonraise d'Entomologie* 23(2): 65–68.

Budiewi E. A. J., Al-Jassany R. & Augul R. S. 2020: Revision of the genus Sinoxylon Duftschmid, 1825 (Coleoptera, Bostrichidae) with new records of species in the middle of Iraq. *Bulletin of the Iraq Natural National History Museum* 16(2): 125–134.

Butze H. 1954: *Im Zweilicht der Tropischen Wälder. Landschaft, Mensch unf Wirtschaft.* Leipzig: F. A. Brockhaus, 446 pp. + 41 figs. + 11 maps.

Cai Ch., Tihelka E., Giacomelli M., Lawrence J. F., Ślipiński A., Kundrata R., Yemamoto S., Thayer M. K., Newton A. F., Leschen R. A., Gimmel M. L., Lü L., Engel M. S., Pouchard P., Hunag D., Pisani D. & Donoghue P. C. J. 2022: Integrated phylogenomics and fossil data illuminate the evolution of beetles. *Royal Society Open Science* 9: 1–86.

Carmen Gerónimo-Torres del J., Pérez-de La Cruz M., Cruz-Pérez de la A., Arias-Rodríguez L. & Burelo-Ramos C. M. 2020: Diversidad y distribución vertical de escarabajos barrenadores (Coleoptera: Bostrichidae, Curculionidae: Scolytinae,

Platypodinae) en un manglar en Tabasco, México. *Caldasia* 43(1): 172–185. doi: https://dx.doi.org/10.15446/caldasia.v43n1.84499.

[1)]Carter H. J. & Zeck E. H. 1937: A Monograf of Australian Colydiidae. *The Proceedings of the Linnean Society of New South Wales* 62(3–4): 181–209.

[1)]Casey T. L. 1885: *Contributions to the Descriptive and Systematic Coleopterology of North America. Part II*. Philadelphia: Colin, pp. 141–198.

Casey T. L. 1890: Coleopterological notices. VII. *Annals of the New York Academy of Science* 5: 307–504.

[1)]Casey T. L. 1891: Coleopterological Notices. III. *Annals of the New York Academy of Science* 6: 9–214.

[1)]Casey T. L. 1898: Studies in the Ptinidae, Ciodidae and Sphindidae of America. *Journal of the New York Entomological Society* 6: 61–93.

[1)]Casey T. L. 1914: Miscellaneous notes and new species. *Memoirs on the Coleoptera* [Lancaster] 5: 355–378.

[1)]Casey T. L. 1924: *Memoirs on the Coleoptera XI*. Lancaster: Lancaster Press, 348 pp.

[1)]Castelnau F. L. N. C. de Laporte, 1836: Études entomologiques, ou descriptions d'Insectes nouveaux et observations sur la synonymie. *Revue Entomologique* [Strabourg, Paris] 4: 5–60.

Cériaco L. M. P., Lima de R. F., Bell R. C. & Melo M. (eds.) 2022: *Biodiversity of the Gulf of Guinea Oceanic Island.* Frankfurt a. M.: Springer, 710 pp.

Chen Z. L. 1990: Note on male of Xylothrips cathaicus Reichardt (Coleoptera: Bostrichidae). *Acta Zootaxonomica Sinica* 15: 255 (in Chinese, with English title).

Chen Z. L. 2000: The Bostrichidae pests intercepted in China (Coleoptera: Bostrichoidea). *Plant Quarantine* 14(3): 153–157 (in Chinese).

Chen Z. L. 2011: *Quarantine and Identification of Bostrichidae (Coleoptera)*. Beijing: China Agriculture Press, 291 pp., 110 pls. (in Chinese)

Chen Z. L. & Yin H. F. 2003: A new species of the genus Dysides from Indonesia (Coleoptera, Bostrichidae). *Acta Zootaxonomica Sinica* 28: 113–115. (in Chinese with English abstract)

Chevrolat L. A. A. 1861: Observations et notes synonymiques. *Annales de la Société Entomologique de France* (4) 1: 389–392.

[1)]Chobaut A. 1897: Description de deux Coléoptères nouveaux du Mzab (Sud-Algérien). *Bulletin de la Société Entomologique de France* 1897: 170–171.

[1)]Chobaut A. 1898: Description de quelques espèces et variétés nouvelles de Coléoptères algériens. *Revue d'Entomologie* [Caen] 17: 74–88.

[1)]Chûjô M. 1936: Description of a new species of Bostrychidae (Coleoptera) and notices of some others. *Transaction of the Natural History Society of Formosa* 26(158): 407–410.

Chûjô M. 1937: *Family Bostrichidae. Family Lyctidae. Class Insecta Coleopteroidea – Coleoptera. Fauna Nipponica. Vol. X, Fasc. VIII, No. VII*. Tokyo, 103 + 2 pp. (in Japanese)

Chûjô M. 1958a: Insects of Micronesia. Coleoptera: Bostrychidae. *Insects of Micronesia* 16(2): 85–104.

Chûjô M. 1958b: Illustration on the Japanese species of beetles belonging to the families Bostrychidae and Lyctidae. *Ageha, Supplementum* 6(1): 1–15.

[1)]Chûjô M. 1964: *Coleoptera from South east Asia III*. Pp. 161–316. In: Kira T. & Umesao T.: Nature and Life South-east Asia. Fauna and Flora Research Society [Kyoto] 3: vii + 466 pp.

[1)]Chûjô M. 1973: An unknown Bostrichidae collected from Tsushima, Japan. *Tôkaishibuhô, Japan Entomological Society* 26: 10. (in Japanese)

Colijn E. O., Beetjens K. K., Butôt R., Miller J. A., Smit J. T., Winter de A. J. & Hoorn van der B. 2019: A catalogue of the Coleoptera of the Dutch Antilles. *Tijdschrift voor Entomologie* 162: 62–186.

[1)]Collart A. 1934: Notes de Chasse au sujet des Coléoptères du Congo Belge. *Bulletin de la Société Entomologique de Belgique* 74: 230–250.

[1)]Comolli A. 1837: *De Coleopteris novis ac rarioribus minusve cognitis provinciae Novocomi*. Ticini: Fusi et Socii, 54 pp.

[1)]Cooper K. W. 1986: A lectotype for Dinapate wrighti Horn, the giant palm-borer and description of a new species of Dinapate from eastern Mexico (Coleoptera, Bostrychidae). *Transaction of the San Diego Society of Natural History* 21: 81–87.

Costa Lima A. M. 1953: *Insetos do Brazil. 8. Tomo. Coleópteros 2ª parte. Escola Nacional de Agronomia. Série didática no. 10*. Rio de Janeiro: Rio de Janeiro: Escola Nacional de Agronomia, 324 pp. + 259 pls.

Curtis J. H. 1839: Descriptions, & c. of the Insects collected by Captain P. P. King, R. N. F. R. S. & L. S. in the Survey of the Straits of Magellan. *The Transactions of the Linnaean Society of London* 18(2): 181–205 + 15 pl.

[3)]Curtis J. H. 1840: Descriptions, & c. of some rare or interesting Indigenous Insects. *Annals of Natural History* 5: 274–282.

Damoiseau R. 1966: Les Bostrychidae du Musée de Moravie à Brno (Coleoptera – Cucujoidea). *Časopis Moravského Musea v Brně* 51: 291–300.

[1)]Damoiseau R. 1968a: Quelques indications nouvelles concernant les Bostrychidae d'Afrique. *Revue de Zoologique et de Botanique Africaines* 77: 303–313.

[1)]Damoiseau R. 1968b: Paraxylogenes, un genre nouveau de Bostrychidae asiatique (Coleoptera – Cucujoidea). *Bulletin de l'Institut Royale des Sciences Naturelles de Belgique* 44(5): 1–6.

[1)]Damoiseau R. 1968c: Resultats scientifiques de l'expedition pedo-zoologique hongroise an Congo-Brazzaville. 24. – Bostrychidae (Coleoptera – Bostrychoidea). *Bulletin de l'Institut Royale des Sciences Naturelles de Belgique* 44(23): 1–6.

Damoiseau R. 1968d: Coleoptera d'Afrique Nord-Orientale. Bostrychidae. 5ème contribution à la connaissance des Bostrychidae. *Notulae Entomologicae* 48: 175–178.

Damoiseau R. 1969a: Contribution a l'faune du Congo (Brazzaville). Mission A. Villiers et A. Descarpentries. 85. Coleopterous. Bostrychidae. *Bulletin de l'Institut Fond. d'Afrique Noire* (Ser. A) 31: 676–679.

Damoiseau R. 1969b: Contribution a la faune de l'Iran. 13. Coléoptères. Bostrychidae. *Annales de la Société Entomologique de France* 5: 143–144.

Damoiseau R. 1973: Coléoptères Bostrichidae d'Afrique orientale. *Monitore Zoologico Italiano,* Supplementum 5: 81–90.

Damoiseau R. 1989: Bostrychoidea. Pp. 234–237. In: Scholtz C. H. & Holm E. (eds.): *Insects of Southern Africa.* Durban: Butterworths, 502 pp.

[1]Damoiseau R. & Coulon G. 1993: Un nouveau Xylopsocus Lesne, 1901 de la region orientale (Coleoptera, Bostrychidae). *Bulletin et Annales de la Société Royale Entomologique de Belgique* 129(1–3): 53–55.

Dejean P. F. M. A. 1802: *Catalogue des coléoptères de la collection d Auguste Dejean.* Paris, 11 pp.

[1], [2]Dejean P. F. M. A. 1821: *Catalogue de la collection des coléoptères de M. le Baron Dejean.* Paris: Crevot, viii-+ 136 + [2] pp.

[1], [2]Dejean P. F. M. A. 1835: *Catalogue de la collection des coléoptères de M. le Comte Dejean. 4ᵉ* Livraison. Paris: Méquignon-Marvis et Fils, pp. 257–360.

Drapiez P. A. J. 1819: Description de huit espèces d'insectes nouveaux. *Annales Générales des Sciences Physiques* 1: 290–298 + pl. 11.

[3]Dufour L. J. M. 1843: Excursion entomologique dans les montagnes de la Vallée d'Ossau. (Catalogue de 767 especes de Coléoptères). *Bulletin de la Société des Sciences, Lettres et Arts, de Pau* 8: 1–118.

[1], [3]Duftschmid C. E. 1825: *Fauna Austriae. Oder Beschreibung der österreichischen Insecten, für angehende Freunde der Entomologie. Dritter Theil.* Linz: Priv. k.k. akademischen Kunst-, Muste- und Buchhandlung, 290 pp.

[1]Erichson W. F. 1834: Coleoptera. In: Erichson W. F. & Burmeister H.: *Beitrage zur Zoologie, gesamelt auf einer Reise um di Erde von Dr. F. J. F. Meyen, M. d. A.d. N. Sechste Abhandlungen. Verhandlungen der Kaiserlichen Leopoldinisch-Carolinischen Akademie der Naturforscher. Achten Bandes Supplement, enthaltend F. J. F. Meyen's Beiträge zur Zoologie, gesammelt auf einer Reise um die Erde, und W. Erichson's und H. Burmeister's Beschreibungen und Abbildungen der von Herrn Meyen auf dieser Reise gesammelten Insecten. Mit 41 Theils Kupfer- Theils Steindrucktafeln.* Breslau und Bonn: Eduard Weber, 219–276 + 41 pls.

[1]Erichson W. F. 1842: Beitrag zur Insecten-Fauna von Vandiemensland, mit besonderer Berücksichtigung der geographischen Verbreitung der Insecten. *Archiv für Naturgeschichte* 8: 83–287 + 2 pl.

[1]Erichson W. F. 1847: Conspectus Insectorum Coleopterorum quae in Republica Peruana observata sunt. *Archiv für Naturgeschichte* 13(1): 67–185.

Español F. 1955: Los bostríquidos de Cataluña y Baleares (Col. Cucujoidea). *Publicaciones del Instituto de Biologia Aplicada* 21: 107–135.

Español F. 1956: Líctidos y bostríquidos de la zona mediterránea de Marruecos. *Publicaciones del Instituto de Biologia Aplicada* 24: 73–75.

Español F. 1959: Algunos Bostriquidos de la Guinea Espanola (Col., Cucujoidea). *Graellsia Revista de Entomólogos Españoles* 17: 53–57.

[1), 3)]Fabricius J. C. 1775: *Systema entomologiae, sistens Insectorum classes, ordines, genera, species, adiectis synonymis, locis, descriptionibus, observationibus*. Flensburgi et Lipsiae: Korte, xxx + 832 pp.

[2), 3)]Fabricius J. C. 1777: *Genera Insectorum eorumque characteres naturales secundum numerum, figuram, situm et proportionem omnium partium oris adiecta mantissa specierum nuper detectarum*. Chilonii: Mich. Friedr. Bartsch, [16] + 310 pp. [1776].

Fabricius J. C. 1781: *Species Insectorum exhibentes eorum differentias specificas, synonyma auctorum, loca natalia, metamorphosin adiectis observationibus, descriptionibus. Tom I.* Hamburgi et Kilonii: C. E. Bohn, viii + 552 pp.

[1), 3)]Fabricius J. C. 1787: *Mantissa insectorum sistens eorum species nuper detectas adiectis characteribus genericis, differentiis specificis, emendationibus, observationibus. Tom I.* Hafniae: C. G. Proft, xx + 348 pp.

[1), 3)]Fabricius J. C. 1792: *Entomologia Systematica emendata et aucta. Secundum classes, ordines, genera, species adjectis synonymis, locis, observationibus, descriptionibus. Tom I. Pars II.* Hafniae: C. G. Proft, 538 pp.

[1), 3)]Fabricius J. C. 1798: *Supplementum entomologiae systematicae*. Hafniae: C.G. Proft et Storch, ii + 572 pp.

[1), 3)]Fabricius J. C. 1801: *Systema Eleutheratorum secundum ordines, genera, species adiectis synonymis, locis, observationibus, descriptionibus. I.* Kiliae: Bibliopolii academici, xxiv + 506 pp.

[1)]Fåhraeus O. I. 1871: Coleoptera Caffrariae, annis 1838–1854 a J. A. Wahlberg collecta. Fam. Scolytidae, Paussidae, Bostrychidae et Cioidae. *Öfversigt af Finska Vetenskaps-Societetens Förhandlingar* 28(6): 661–672.

[1)]Fairmaire L. 1850: Essai sur les Coléoptères de la Polynésie. *Revue et Magasin de Zoologie* 2(2): 50–64.

[1)]Fairmaire L. 1874: Note sur quelques Coléoptères des environs de Tuggurth. *Petites Nouveles Entomologiques* 6: 407–408.

[1)]Fairmaire L. 1880a: Diagnoses de Coléoptères de Nossi-Bé. *Le Naturaliste, Revue Illustrée des Sciences Naturelles* 2: 246.

[1)]Fairmaire L. 1880b: Diagnoses de Coléoptères de Madagascar. *Le Naturaliste, Revue Illustrée des Sciences Naturelles* 2: 307–308.

Fairmaire L. 1882: Notes sur quelques Coléoptères du Soudan et de l'Inde boréale recueillis par MM. Stanislas et Constantin Rembielinski. *Annales de la Société Entomologique de France* (6) 2: 65–68.

[1)]Fairmaire L. 1883a: Descriptions de Coléoptères nouveaux ou peu connus récoltés par Mr. Raffray en Abyssinie. *Annales de la Société Entomologiques de France* (6) 3: 89–112.

[1)]Fairmaire L. 1883b: Descriptions de deux espèces nouvelles de Coléoptères africains. *Bulletin de la Sociéte Entomologique de France* 1883: cxxxiii–cxxxiv.

[1)]Fairmaire L. 1883c: Diagnoses de Coléoptères Abyssins. *Le Naturaliste, Revue Illustrée des Sciences Naturelles* 5: 205–206.

[1)]Fairmaire L. 1888: Énumeration des Coléoptères recueillis par M. le Dr. Hans Schinz dans le sud de l'Afrique. *Annales de la Sociéte Entomologique de France* (6) 8: 173–202.

[1)]Fairmaire L. 1892a: Coléoptères d'Obock. Troisième partie. *Revue d'Entomologie* 11: 77–127.

[1)]Fairmaire L. 1892b: Descriptions de quelques Coléoptères argentins. *Annales de la Société Entomologique de Belgique* 36: 242–253.

[1)]Fairmaire L. 1893a: Note sur les Coléoptères du Choa. *Annales de la Société Entomologique de Belgique* 37: 9–50.

Fairmaire L. 1893b: Coléoptères des iles Comores. *Annales de la Société Entomologique de Belgique* 37: 521–555.

[1)]Fairmaire L. & Germain P. 1861: *Coleoptera Chilensia*. 2. Paris: 8 + 8 pp.

[1)]Faldermann F. 1837: Fauna entomologica Trans-caucasica. Coleoptera. Pars II. *Nouveaux Mémoires de la Société Impériale des Naturalistes de Moscou* 5: 1–433 pp. + 15 pls.

[1)]Fall H. C. 1901: List of; the Coleoptera of Southern California, with Notes on Habits and Distribution and Description of New Species. *Occasional Papers of the California Academy of Sciences* 8: 1–282.

[1)]Fall H. C. 1909: New Coleoptera from the South-West. – IV. *The Canadian Entomologist* 41(5): 161–170.

Farashiani M. E., Varandi H. B., Kazerani F., Yarmand H., Babaee M., Thorn S., Lange F., Rafiei-Jahed R., Müller J. & Amini S. 2022: A preliminary checklist of saproxylic beetles (Coleoptera) in the Hyrcanian forests of Iran, with distributional data. *Check List* 18(5): 1063–1120.

[1)]Fauvel A. 1904: Faune analytique des Coléoptères de la Nouvelle-Caledonie. 2e partie. *Revue d'Entomologie* 23: 113–207.

Ferrú M. & Elgueta M. 2011: Lista de coleoptéros (Insecta: Coleoptera) de la regiones de Arica y Paranicota y de Tarapacá, Chile. *Boletín del Museo Nacional de Historia Natural* (Chile) 62: 9–61.

[1)]Fisher W. S. 1950: A revision of the North American species of beetles belonging to the family Bostrychidae. *Miscellaneous Publications. United States Department of Agriculture*, Washington, No. 698: 1–157.

[1)]Forbes T. M. 1926: The Wing Folding Patterns of the Coleoptera (Continued). *Journal of the New York Entomological Society* 34(2): 91–139.

Franz H. 1985: Beitrag zur Kenntnis der Koleopterenfauna der Galapago-Inseln. *Sitzungsberichte der Österreichischen Akademie der Wissenschaften. Mathematical-naturwissinschaftliche Klasse. Abteilung I.* 194: 73–124.

Franz H. 1987: Biogeographische und ökologische Studien auf den Kapverdischen Inseln. Sitzungsberichte der Österreichischen Akademie der Wissenschaften. *Mathematical-naturwissinschaftliche Klasse.* Abteilung I. 196: 89–110.

[1)]Frivaldszky von Frivald E. 1835: Közlések a Balkány vidékén tett Természettudományi utazásrol. *A Magyar Tudós Társaság Évkönyvei* 2: 235–276 + 7 pls.

[1)]Fursov N. I. 1936: Die neuen Arten der Holzkäfer (zwei aus Fam. Eucnemidae aus Ost-Siberien und eine aus Fam. Bostrichidae aus Mittelasien). *Bulletin de la Société des Naturalistes de Moscou* (N. S.) 45(5): 348–350. (in Russian)

[1)]Fursov N. I. 1939: Eine neue Gattung und 6 neue Colepterenarten von Mittelasien und Caucasus. *Bulletin de la Société des Naturalistes de Moscou* (N. S.) 48(1): 88–90.

[1)]Galeazzi G. 1854: *Coleoptera Europae dubleta, quae pro mutica commutatione offeri possunt.* Milano: 19 pp.

[1)]Gardner J. C. M. 1933: Immature stages of Indian Coleoptera (13). Bostrychidae. *Indian Forest Record, Entomology Series* 18(9): 1–19 + 4 pls.

Geis K.-U. 2002: Gebietsfremde Splintholz- und Bohrkäfer, nach Mitteleuropa mit Importholz und anderen Gütern eingeschleppt. Eine Bestandsaufnahme (Coleoptera: Lyctidae, Bostrichidae). *Mitteilungen des Internationalen Entomologischen Vereins e. V.* Supplement 10: 1–106.

[1)]Geis K.-U. 2015: Neue Beiträge zur Fauna der Bostrichidae (Coleoptera) der Arabischen Halbinsel und Sokotras. *Mitteilungen des Internationalen Entomologischen Vereins* 40: 63–101.

Geisthardt M. 2010a: Zur Kenntnis der Bohr- und Splintholzkäfer des Jemen. *Mitteilungen des Internationalen Entomologischen Vereins e. V.* 35(1/2): 29–42.

Geisthardt M. 2010b: Order Coleoptera, family Bostrichidae. *Arthropods Fauna of UAE* 3: 204–225.

[1)]Gemminger M. & Harold E. 1869: *Catalogus Coleopterorum hucusque descriptorum synonymicus et systematicus. Tomus VI. Rhipidoceridae, Dascillidae, Malacodermidae, Cleridae, Lymexylonidae, Cupesidae, Ptinidae, Bostrychidae, Cioidae.* Monachii: E. H. Gummi, 1609–1800 + 5 pp.

[1)]Geoffroy E. L. 1762: *Histoire abrégée des Insectes qui se trouvent aux environs de Paris; Dans laquelle ces Animaux sont rangés suivant un ordre méthodique. Tome premier.* Paris: Durand, xxviii + 523 pp. + 10 pls.

[1)]Geoffroy E. L. 1785: [new taxa] In: Fourcroy A. F. de: *Entomologia parisiensis; sive Catalogus Insectorum quae in Agro Parisiensi reperiuntur; Secundum methodum Geoffraeanam in sectiones, genera et species distributus: cui addita sunt nomina trivialia et fere trecentae novae Species. Pars prima.* Parisiis: Privilegio Academiae, vii + [1] + 231 pp.

[3)]Georg W. 1856: Bostrichus Alni, ein neue entdecker Käfer (nebst Nachschrift von Ratzeburg). *Entomologische Zeitung* [Stettin] 17(1–2): 59–60.

[1)]Gerberg E. J. 1957: A Revision of The New World Species of Powder-Post Beetles Belonging to the Family Lyctidae. *Technical Bulletin. United States Department of Agriculture* (1157): 1–55 + XIV pl.

[1)]Germar E. F. 1817: *Reise nach Dalmatien und in das Gebiet von Ragusa*. Leipzig und Altenburg: F. A. Brockhaus, xii + 323 pp. + 9 pls. + 2 maps.

[1)]Germar E. F. 1824: *Insectorum species novae aut minus cognitae, descriptionibus illustratae. Volumen primum. Coleoptera*. Halae: J. C. Hendel et filii, xxiv + 624 pp. + 2 pls.

[1)]Germar E. F. 1848: Beiträge zur Insectenfauna von Adelaide. *Linnaea Entomologica* 3: 153–247.

[1)]Gerstäcker C. E. A. 1855: Diagnosen der von ihm in Mossambique gesammelten Käfer aus den Familien der Longicornia, Paussidae und Ptinoires. *Monatsbericht der Königliche Akademie der Wissenschaften zu Berlin* 1855: 265–268.

[1)]Gestro E. 1895: Esplorazione del Giuba e dei suoi affluenti computa dal Cap. V. Bottego durante gli anni 1892–93 sotto gli auspicii delta Società Geografica Italiana. Risultati Zoologici. Coleoptera. *Annali del Museo Civico di Storia Naturale di Genova* (*Série 2*) 15(35): 255–478.

Gino J. N., Uzbekia G. C. & Háva J. 2015: First record of Dolichobostrichus angustus Steinheil, 1872 (Coleoptera: Bostrichidae) for Peru. *The Biologist* [Lima] 13(2): 437–441.

[1)]Giorna M. S. 1792: *Calendario entomologico ossia osservazioni sulla stagioni proprie agl'insetti nell clima Piemontese, e particolarmente ne'contorni di Torino*. Torino: Nella Stamperia Reale, 146 pp.

[3)]Gistel J. N. F. X. 1831: Entomologische Fragmenta. *Isis von Oken* 1831 (3): 301–310.

[1)]Gistel J. N. F. X. 1848: *Naturgeschichte des Thierreichs, für höhere Schulen. Mit einem Atlas von 32 Tafeln (darstellend 617 illuminirte Figuren) und mehreren dem Texte eingedruckten Xylographien*. Stuttgart: Hoffmann, xvi + 216 + [4] pp. + 32 pls.

[2)]Gistel J. N. F. X. 1856: *Die Mysterien der europäischen Insecten Welt*. Kempten: T. Dannheimer, xii + 532 pp.

[2)]Gistel J. N. F. X. 1857: Achthundert und zwanzig neue oder unbeschriebene wirbellose Thiere. Straubing: Schorner'schen Buchhandlung, pp. 513–606.

[2)]Gmelin J. F. 1790: *Caroli Linne Systema Naturae per regna tria naturae, secundum classes, ordines, genera, et species cum characteribus differentis, synonymis, locis. 1*(*4*). *Edition 13*. Lipsiae: Beer, 1517–2224.

[1)]Goeze J. A. E. 1777: *Entomologische Beyträge zu des Ritter Linné. Zwölfter Ausgabe des Natursystems. Erster Theil*. Leipzig: Weidmanns Erben und Reich, xvi + 736 pp.

Gomy Y., Lamagnen R. & Poussereau J. 2016: *Les Coléoptères de l'île de la Réunion*. Saint-Denis: Orphie, 760 pp.

[1)]Gorham H. S. 1883: *Insecta. Coleoptera. Vol. III. Part 2. Malacodermata*. Pp. 169–224. In: Godman F. D. & Salvin O. (eds): Biologia Centrali-America 3 (2), 1880–1886. London: Dulau & Co., B. Quaritsch, xii + 372 + 13 pl.

[1)]Gorham H. S. 1885: Description of a new genus of Bostrychidae. *Notes from the Leyden Museum* 7: 51–53.

[1)]Gorham H. S. 1886: *Insecta. Coleoptera. Vol. III. Part 2. Supplement to Malacodermata*. Pp. xii + 313–372. In: Godman F. D. & Salvin O. (eds): Biologia Centrali-America 3(2), 1880–1886. London: Dulau & Co., B. Quaritsch, xii + 372 + 13 pl.

[1)]Gozis M. des 1881: Quelques rectifications synonymiques touchant différents genres et espèces de Coléoptères français (5e partie). *Bulletin des Séances de la Société Entomologique de France* 1881: 134–135.

[3)]Gravenhorst J. L. C. 1807: *Vergleichede Übersicht des Linneischen und einiger neuern zoologischen Scheme. Nebst dem eingeschalteten Verzeichnisse der zoologischen Sammlung des Verfassers und den Beschreibungen neuer Thierarten, die in derselben vorhanden sind*. Gottingen: Heindrich Dieterich, xx + 476 pp.

[2)]Grimmer K. H. B. 1841: *Steiermark's Coleoptern mit Einhundert sechs neu beschreibenen Species*. Grätz: C. Tanzer, iv + 5–50 pp.

[1)]Grouvelle A. 1896: Nitidulides, Colydiides, Cucujides et Parnides récoltés par M. E. Gounelle au Brésil et autres Clavicornes nouveaux d'Amérique. *Annales de la Société Entomologique de France* 65: 177–216.

Grouvelle A. 1900: Clavicornes. Pp. 424–425. In: Insectes du Congo. Clavicornes. (Récoltés faites a Kinchassa par M. Waelbroeck). *Annales de la Société Entomologique de Belgique* 44: 423–435.

[1)]Guérin-Méneville F. E. 1844: *Iconographie du Règne Animal de G. Cuvier, ou représentation d'après nature de l'une des espèces les plus remarquables, et souvent non encore figurées, de chaque genre d'animaux. VII. Insectes*. Paris: J. B. Baillière, 576 pp. + 104 pl.

[1)]Guérin-Méneville F. E. 1845: Notes sur quelques coléoptères trouvés dans la racine de Squine (Smilax chine). *Bulletin de la Société de Entomologique de France* (2) 3: 16–17.

[3)]Guyon G. 1855: Occurrence at Richmond, Surrey, of a Coleopterous Insect new to Britain. *The Zoologist* 13: 4815.

[3)]Gyllenhal L. 1813: *Insecta Suecica descripta. Classis I. Coleoptera sive Eleutherata. Pars III*. Scaris: Leverentz, ii + 730 pp.

[3)]Gyllenhal L. 1827: *Insecta Suecica descripta. Classis I. Coleoptera sive Eleuterata. Tom. I. Pars IV*. Lipsiae: F. Fleischer, xi + 761 pp.

Hagstrum D. W. & Subramanyam B. 2009: *Stored-product Insect resource*. St. Paul: AACC International, viii + 509 pp.

Halperin J. & Damoiseau R. 1980: The bostrychid beetles (Coleoptera) of Israel. *Israel Journal of Entomology* 14: 47–53.

[3]Hartig G. L. & Hartig T. 1834: *Forstliches und forstnaturwissenschatliches Conversations-Lexicon. Ein Handbuch für Jeden, der sich für das Forstwessen und die dazu gehdringen Naturwissenschtten interesiert. Mit allerhöchsten Privilegien gegen den Nachdruck und den Verkauf desselben.* Berlin: Nauck, xiv + 1034 pp. + [2] pp.

Hartig G. L. & Hartig T. 1836: *Forstliches und forstnaturwissenschatliches Conversations-Lexicon. Ein Handbuch für Jeden, der sich für das Forstwessen und die dazu gehdringen Naturwissenschtten interesiert. Mit allerhöchsten Privilegien gegen den Nachdruck und den Verkauf desselben. Zweite, revidirte Auflage* Berlin: Nauck, xiv + 1034 pp. + [2] pp.

Háva J. 2017: New genus and species for Lebanon, Psoa dubia (Rossi, 1792) (Coleoptera: Bostrichidae: Psoinae). *Arquivos Entomolóxicos* 17: 455–456.

Háva J., Bureš L. & Kopr D. 2022: The families Dermestidae and Bostrichidae (Coleoptera: Bostrichoidea) in the United Arab Emirates. *Euroasian Entomological Journal* 21(6): 358–360.

Háva J. & Chaboo C. S. 2015: Beetles (Coleoptera) of Peru: A Survey of the Families. Nosodendridae (Derodontoidea), Dermestidae, Bostrichidae (Bostrichoidea). *Journal of the Kansas Entomological Society* 88(3): 404–407.

Háva J. & Legalov A. 2023a: A new Poinarinius species (Coleoptera: Bostrichidae: Alitrepaninae) from mid-Cretaceous Burmese amber. *Studies and Reports, Taxonomical Series* 19(2): 285–287.

Háva J. & Legalov A. 2023b: Poinarinius coziki sp. n., a new species of bostrychid beetles (Coleoptera: Boistrichidae) from Cretaceous Burmese amber. *Euroasian Entomological Journal* 22(5): 272–273.

Háva J. & Zahradník P. 2021: New faunistic records of Apate bicolor Fåhreus, 1781 from the Afrotropical Region (Coleoptera: Bostrichidae: Apatinae). *Munis Entomology* 16(2): 830–832.

[3]Heer O. 1836: *Observationes entomologicae, continentes metamorphoses Coleopterorum nonnulorum adhuc incognitas.* Turici: Orellii, Fuesslini & Sociorum, [1] + 36 pp. + 6 pls.

Hellwig J. C. L. 1792: Dritte Nachricht von neuen Gattungen in entomologischen System. *Neuestes Magazin für die Liebhaber der Entomologie* 1 (3): 385–408.

Hendrych R. 1984: Fytogeografie. [Phytogeography]. Praha: Státní pedagogické nakladatelství, 221 pp.

[1], [3]Herbst J. F. W. 1783: Kritisches Verzeichniss meiner Insektensammlung. *Archiv der Insectengeschichte* [Zürich: J. C. Füessly] 4 (1): 1–72 + pls. 19–24.

[3]Herbst J. F. W. 1784a: Kritisches Verzeichniss meiner Insektensammlung. *Archiv der Insectengeschichte* [Zürich: J. C. Füessly] 5 (1): 73–128 + pls. 24–28.

[3]Herbst J. F. W. 1784b: Kritisches Verzeichniss meiner Insektensammlung. *Archiv der Insectengeschichte* [Zürich: J. C. Füessly] 5 (2): 129–151 + pls. 29–30.

Herbst J. F. W. 1793: *Natursystem aller bekannten in- und ausländischen Insecten, als eine Fortsetzung der von Büffonschen Naturgeschichte. Der Käfer fünfter Theil.* Berlin: Pauli, 392 pp. + 16 pl.

[1)]Herbst J. F. W. 1797: *Natursystem aller bekannten in- und ausländischen Insecten, als eine Fortsetzung der von Büffonschen Naturgeschichte. Der Käfer siebenter Theil.* Berlin: J. Pauli, 346 pp. + 26 pl.

Heyden L. F. J. D. von, Reitter E. & Weise J. 1883: *Catalogus Coleopterorum Europae et Caucasi. Editio tertia.* Londini, Berolini, Parisiis: E. Janson, Libraria Nicolai, L. Buquet, 2 + 228 pp.

[1)]Heyne A. & Taschenberg O. 1907: *Die Exotischen Käfer in Wort und Bild. Band I. Text.* Leipzig: G. Reusche, (23–24): 195–218.

[1)]Hope F. W. 1845: On the Entomology of China, with Descriptions of the new Species sent to England by Dr. Cantor from Chusan and Canton. *The Transaction of the Entomological Society of London* 4: 4–17.

Hopkins A. D. 1911: Notes on habits and distribution, with list of described species. Appendix. *United States Department of Agriculture Entomology, Technical Bulletin* 20: 130–138.

[1)]Horn G. H. 1868: New species of Coleoptera from the Pacific District of the United States. *Transactions of the American Entomological Society* 2: 129–140.

[1)]Horn G. H. 1878: Revision of the Species of the Sub-family Bostrichidae of the United States. *Proceedings of the American Philosophical Society Held at Philadelphia for Promoting Useful Knowledge* 17: 540–555.

[1)]Horn G. H. 1885: Contribution to the Coleopterology of the United States. *Transactions of the American Entomological Society and Proceedings of the Entomological Section of the Academy of Natural Sciences* 12: 128–162.

[1)]Horn G. H. 1886: Dinapate Wrightii and its larva. *Transactions of the American Entomological Society* 13: 1–4.

Horn G. H. 1894: The Coleoptera of Baja California. *Proceedings of the California Academy of Sciences* (2) 4: 302–449 + 2 pl.

Horn G. H. 1896: The Coleoptera of Baja California. (Supplement II.). *Proceedings of the California Academy of Sciences* (2) 6: 367–381.

Horn W., Kahle I., Friese G. & Gaedike R. 1990a: *Collectiones Entomologicae. Teil I: A bis K.* Berlin: Akademie der Landwirtschftenwissenschaften der Deutschen Demokratischen Republik, 1–220 pp.

Horn W., Kahle I., Friese G. & Gaedike R. 1990b: *Collectiones Entomologicae. Teil II: L bis Z.* Berlin: Akademie der Landwirtschftenwissenschaften der Deutschen Demokratischen Republik, 221–573 pp.

[3)]Hornung E. G. 1842: Ueber einige in den Betelnüssen vorkommende Käfer. *Entomologische Zeitung* [Stettin] 3(5): 115–117.

[1)]Hua L.-Z. 2002: *List of Chinese Insects. Vol. II.* Guangzhou China: Zhongshan (Sun Yat-sen) University Press, 612 pp.

[1])Illiger J. K. W. 1801: Neue Insecten. *Magazin für Insektenkunde* [Braunschweig] 1/1.2.: 163–208.

[1])Iablokoff-Khnzorian S. M. 1976: The powder-post beetles of USSR (Coleoptera, Lyctidae). *Zoologicheskiy Sbornik* [Yerevan] 17: 87–100. (in Russian).

ICZN (International Commission on Zoological Nomenclature) 1999: *International Code of Zoological Nomenclature. Fourth edition adopted by the International Union of Biological Sciences*. London: The International Trust for Zoological Nomenclatura, i–xxix + 306 pp.

[1])Imhoff L. 1843: Beschreibung einer Anzahl Guineensicher Käfer vom Missionair Riis. *Bericht über die Verhandlungen der Naturforschenden Geselschaft in Basel* 5: 164–180.

Ivie M. A. 2010: Additions and corrections to Borowski and Węgrzynowicz's world catalogue of Bostrichidae (Coleoptera). *Zootaxa* 2498: 28–46.

Jacobi A 1906: *Grundriss der Zoologie für Forstleute. Ergänzungsband zu Lorey's Handbuch der Forstwissenschaft*. Tübingen: H. Laupp, xi + 263 pp.

[1])Jacobson (= Jakobson) G. G. 1915: *Zhuki Rossii i zapadnoy Evropy. [Beetles of the Russian and the West Europa]*. St. Peterburg: Izd. A. F. Devriena, 865–1024 pp. (in Russian).

[1])Jacquelin Du Val P. N. C. 1859: Description de quelques espèces nouvelles. *Glanures Entomologiques* 1: 34–42.

Jacquelin Du Val P. N. C. 1860: Remarques et synonymies diverses. *Glanures Entomologiques* 2: 161–164.

[1])Jacquelin Du Val P. N. C. 1861: *Manual entomologique. Genera des Coléoptères d'Europe comprenant leur Clasification en Familles naturelles, la Description de tous les Genres, des Tableaux synoptiques destinés à faciliter l'étude, le Catalogue de Toutes les espececes, de nombreux dessins au trait de Caractéres. Tome 3.–2e Partie. Buprestides, Throscides, Eucnemides, Élaterides, Cébrionides, Rhipicérides, Dascillides, Lampyrides, Téléphoridés, Malachiides, Clérides, Lyméxylonides, Ptinides, Anobiides, Sphindides, Apatides, Lyctides, Cisides*. Paris: A. Deyrolle, 89–240 + 139–168 pp. + pls. 22–58.

[1])Karsch F. A. F. 1881: Die Käfer der Rohlfs'schen Afrikanischen Expedition 1878–79. *Berliner Entomologische Zeitschrift* 25: 41–50 + 2 pl.

[1])Kiesenwetter E. A. H. von 1877: Anobiidae und Cioidae. Pp. 1–200. In: Kiesenwetter E. A. H. von & Seidlitz G. 1877–1898: *Naturgeschichte der Insecten Deutschlands. Erste Abtheilung. Coleoptera. Fünfter Band. Erste Hälfte*. Berlin: Nicolaische Verlags-Buchhandlung R. Stricker, xxviii + 877 pp.

Kiesenwetter E. A. H. von 1879: Coleoptera Japoniae collecta a Domino Lewis et aliis. *Deutsche Entomologische Zeitschrift* 23: 305–320.

[3])Kirby W. 1837: Part the fourth and last. The Insects. In: Richardson J. (ed.): *Fauna Boreali-Americana, or the Zoology of the northern parts of British America: containing descriptions of the objects of natural history collected on the late northern land expeditions, under command of Captain Sir John Franklin, R. N.* Norwich: J. Fletcher, xxxix + 325 pp. + 2 p, 8 pls.

[3]Kirsch T. F. W. 1871: Beschreibung des Bostrichus (Tomicus) Judeichii n. sp. *Berliner Entomologische Zeitschrift* 14(4) [1870]: 388.

[1]Klug J. C. F. 1833: Bericht über eine auf Madagascar veranstaltete Sammlung von Insecten aus der Ordnung Coleoptera. *Abhandlungen der Königlischen Akademie der Wissenschaften, Physikalisch-Matematische Klasse* 19[1832–1833]: 91–223 + 5 pls.

Knížek M. 2011: Family Curculioniade, subfamilies Platypodinae, Scolytinae. Pp. 201–251. In: Löbla I. & Smetana A. (eds.): *Catalogue of Palaearctic Coleoptera. Curculionoidea I. Volume 7*. Stenstrup: Apollo Books, 373 pp.

Kojima T. 1932: Beiträge zur Kenntnis von Lyctus linearis Goeze. *Zeitschrift für Angewandte Entomologie* 19(3): 325–356.

[1]Kocher L. 1956: Catalogue commenté des Coléoptères du Maroc. Fasc.IV. Clavicornes et groupes voisins. *Travaux de l'Institut Scientifique Chérifien, Série Zoologie* 11: 1–136.

[3]Kolenati F. A. 1846: *Meletemata entomologica. Brachelytra Caucasi cum distributione geographica adnexis pselaphinis, scydmaenis, notoxidibus et xylophagis. Fasc. III. Accedunt tabulae III coloratae*. Petropoli: Imperialis Academiae Scientarium, [3] + 44 pp. + pls 12–14.

[1]Kraus E. J. 1911: Technical papers on miscellaneous forest Insects. III. A revision of the powder-post beetles of the family Lyctidae of The United States and Europe. *United States Department of Agriculture Entomology. Technical Bulletin* 20: 111–129.

[3]Küster H. C. 1845: *Die Käfer Europa's. Nach der Natur beschrieben. Zweites Heft*. Nürnberg: von Bauer & Raspe, [6] + 100 sheets + 2 pls.

[1]Küster H. C. 1847: *Die Käfer Europa's. Nach der Natur beschrieben. Neuntes Heft*. Nürnberg: von Bauer & Raspe, [4] + 100 sheets + 3 pls.

[1], [2], [3]Kugelann J. G. 1792: Verzeichniss der in einigen Gegenden Preussens bis jetzt entdeckten Käfer-arten, nebst kurzen Nachrichten von denselben. *Neustes Magazin für die Liebhaber der Entomologie* 1(4): 477–512.

[3]Kugelann J. G. 1794: Verzeichniss der in einigen Gegenden Preussens bis jetzt entdeckten Käfer-arten, nebst kurzen Nachrichten von denselben. *Neustes Magazin für die Liebhaber der Entomologie* 1(5): 513–582.

[3]Lacepède B. G. E. 1802: *Histoire naturelle des poissons. Tome troisieme*. Paris: Plassan [1801] i–lxvi + 1–558, Pls. 1–34.

Lacordaire J. T. 1857: *Histoire Naturelle des Insectes. Genera des Coléoptères ou exposé méthodique et critique de tous les genres proposés jusqu'ici dans cet ordre d'Insectes. Tome Quatrième contenant les familles des Buprestides, Throscides, Eucnémides, Élateriders, Cébrionides, Céropythides, Rhipicérides, Dascillides, Malacodermes, Clérides, Lyméxylones, Cupésides, Ptiniores, Bostrichides et Cissides*. Paris: Roret, 580 pp.

Laicharting J. N. von 1781: *Verzeichnis und Beschreibung der Tyroler-Insecten. I. Theil. Käferartige Insecten*. Zürich: J. C. Füessly, [2] xii + 248 pp.

[1)]Latreille P. A. 1802: *Histoire naturelle, générale et particulière, des Crustacés et des Insectes. Ouvrage faisant suite aux oeuvres de Leclerc de Buffon, et partie du cours complet d'histoire naterelle rédigé par C. S. Sonnini, membre de plusiers sociétés savantes. Familles naturelles des genres. Tome troisieme.* Paris: F. Dufart, xii + 13–467 pp. + [1 p. errata].

[1)]Latreille P. A. 1807: *Genera Crustaceorum et Insectorum secundum ordinem naturalem in familias disposita, iconibus exemplisque plurimis explicata. Tomus tertius.* Parisiis et Argentorati: A. Koenig, 258 pp.

[1)]Latreille P. A. 1813: *Insectes de l'Amérique équinoxiale, recueillis pendant le voyage de MM. De Humboldtet Bonpland.* In: *Humboldt A. & Bonpland A.: Voyage de Humbolt et Bonpland. Deuxième Partie. Observations de Zoologie et d'anatomie comparée, faite dans l'Océan atlantique, dans l'intérieur su Nouveau Continent et dans la Mer du Sud, pendant les années 1799, 1800, 1801, 1802 et 1803. Seconde volume.* Paris: F. Schoell et G. Dufour et comp., 138 pp. + 13 pl.

[1)]Latreille P. A., Le Peletier de Saint-Fargeau A., Serville J. G. A. & Guérin-Méneville F. E. 1825: *Histoire Naturelle. Entomologie, ou Histoire Naturelle des Crustacés, des Arachnides et des Insectes. Tome Dixième. Par – Sca.* In: Latreille P. A. (ed.): Encyclopédie méthodique. Paris: Agasse, 7 + 344 pp.

Lawrence J. F. 1980: A new genus of Indo-Australian Gempylodini with notes on the constitution of the Colydidae (Coleoptera). *Journal of the Australian Entomological Society* 19: 293–310.

[1)]Lawrence J. F. & Ślipiński A. 2013: Loranthophila, a new genus of Australian Lyctinae (Coleoptera: Bostrichidae) associated with Mistletoe. *Zootaxa* 3737(3): 295–300.

[1)]Lea A. M. 1894: Descriptions of new species of Bostrychidae. *The Proceedings of the Linnean Society of New South Wales* (2) 8 [1893]: 317–323.

[1)]LeConte J. L. 1852: Descriptions of new species of Coleoptera, from California. *Annals Lyceum Natural History New York* 5: 125–216.

[1)]LeConte J. L. 1853: Descriptions of twenty new species of Coleoptera inhabiting the United States. *Proceedings of the Academy of Natural Science of Philadelphia* 6: 226–235.

[1)]LeConte J. L. 1858: Description of New Species of Coleoptera, chiefly collected by the United States and Mexican Boundary Commission, under Major W. H. Emory, U.S.A. *Proceedings of the Academy of Natural Sciences of Philadelphia* 10: 59–89.

[1)]LeConte J. L. 1859: Reports of explorations and surveys for a railroad route from the Mississippi River to the Pacific Ocean. 9. 1. *Report upon Insects collected on the Survey.* 1860 (Sept. 1857) 4: 1–72 + 2 pls.

[1)]LeConte J. L. 1861: *Classification of the Coleoptera of North America. Prepared for the Smithsonian Institution. Part 1.* 1. Washington: Smithsonian Institute, xxiv + 214 pp.

[1)]LeConte J. L. 1862: *Classification of the Coleoptera of North America. Prepared for the Smithsonian Institution. Part 1.* 2. Washington: Smithsonian Institute, 209–286 pp.

[1)]LeConte J. L. 1866: *New species of north American Coleoptera. Prepared for the Smithsonian Institution. Part 1. Smithsonian Miscellaneous Collections 167. Part 1.* Washington: Smithsonian Institute: 87–177 pp.

[1)]LeConte J. L. 1868: Coleoptera of the U. S. Coast Survey expedition to Alaska, under charge of Mr. George Davidson. *Transactions of the American Entomological Society* 2: 59–64.

[1)]LeConte J. L. 1874: Descriptions of New Coleoptera chiefly from the Pacific Slope of North America. *Transaction of the American Entomological Society* 5: 43–72.

Leech H. B. 1958: Synonymy of Dinoderus pubicollis Van Dyke (Coleoptera, Bostrychidae). *The Pan-Pacific Entomologist* 34: 230–21.

[1)]Legalov, A. A. 2018: New auger beetle (Coleoptera; Bostrichidae) from mid-Cretaceous Burmese amber. *Cretaceous Research* 92: 210–213.

[1)]Legalov A. A. & Háva J. 2020: The first record of subfamily Polycaoninae (Coleoptera; Bostrichidae) from mid-Cretaceous Burmese amber. *Cretaceous Research* 116: 1–5. https://doi.org/10.1016/j.cretres.2020.104620.

[1)]Legalov A. A. & Háva J. 2022: Diversity of Auger Beetles (Coleoptera: Bostrichidae) in the Mid-Cretaceous Forests with Description of Seven New Species. *Diversity* 2022, 14, 1114. https://doi.org/10.3390/d14121114.

Leng C. W. 1920: *Catalogue of the Coleoptera of America north of Mexico. Mt. Vernon.* N. York: J. D. Sherman, 470 pp.

[1)]Lesne P. 1894: Le genre Dysides Pert. (Apoleon Gorh. in part). *Annales de la Société Entomologique de France* 63: 18–21.

[1)]Lesne P. 1895a: Descriptions de genres nouveaux et d'espèces nouvelles de Coléoptères de la famille des Bostrychides. *Annales de la Société Entomologique de France* 64: 169–178.

[1)]Lesne P. 1895b: Note sur trois Coléoptères de la famille des Bostrychides (Xylopertha dominicana Fabr., X. religiosa Boisd., Bostrychus Künckeli n. sp.) (Col.). *Bulletin de la Société Entomologique de France* 64: 172–179.

[1)]Lesne P. 1896a: Révision des Coléoptères de la famille des Bostrychides. 1er Mémoire. Bostrychides procéphales. *Annales de la Société Entomologique de France* 65: 95–127 + 2 pl.

[1)]Lesne P. 1896b: Contributions à la faune indo-chinoise. 17e Mémoire. Bostrychidae. *Annales de la Société Entomologique de France* 65: 511–512.

Lesne P. 1896c: Notes synonimiques sur les Bostrychides hypocéphales [Col.]. *Bulletin de la Société Entomologique de France* 65: 334–336.

Lesne P. 1897a: Descriptions de deux espèces nouvelles de Bostrychides algériens [Col.]. *Bulletin de la Société Entomologique de France* 1897: 235–237.

[1)]Lesne P. 1897b: Bostrychides indiens de la collection H.-E. Andrewes. *Annales de la Société Entomologique de Belgique* 41: 18–22.

[1])Lesne P. 1897c: Sur une espèce nouvelle de Coléoptère de la famille des Bostrychides (Heterarthron subdepressus n. sp.). *Bulletin de la Société Entomologique de France* 1897: 146–147.

Lesne P. 1897d: Rectification synonymiques relatives au genre Dinoderus (Bostrychidae) [Col.]. *Bulletin de la Société Entomologique de France* 1897: 147.

[1])Lesne P. 1897e: Description d'une espèce nouvelle de Dinoderus. (Coleoptera: Bostrychidae). *Notes from the Leyden Museum* 19: 184.

[1])Lesne P. 1898a: Note sur une espéce nouvelle de Bostrychus (Coleoptera: Bostrychidae). *Notes from the Leyden Museum* 20: 255.

Lesne P. 1898b: Sur un Coléoptère nouveau de la famille des Lyctides. *Bulletin du Muséum d'Histoire Naturelle* 4: 139–140.

[1])Lesne P. 1898c: Revision des Coléoptères de la famille des Bostrichides. 2e Mémoire. Bostrychides hypocéphales. Dinoderinae. *Annales de la Société Entomologique de France* 66: 319–350.

[1])Lesne P. 1899a: Révision des Coléoptères de la famille des Bostrychides. 3e Mémoire. Bostrychinae. Bostrichinae sens. strct. – I. Les Bostrychus. *Annales de la Société Entomologique de France* 67: 438–621.

[1])Lesne P. 1899b: Liste des Bostrychides et Lyctide recueillis sur le littoral de la baie de Tadjourah et description d'une espèce nouvelle. *Bulletin du Muséum d'Histoire Naturelle* 5: 226–229.

Lesne P. 1899c: Bostrychidae. Pp. 10–11. In: Liste des Coléoptères recueillis à Madagascar par MM. le commandant Dorr, de l'Infanterie de marine (1896–97), et le lieutenant Jobit, du 13e d'Artillerie (1895–96). *Mémoires de la Société Zoologique de France* 12: 5–28.

[1])Lesne P. 1899d: Liste de Bostrychides des collections du Musée Civique de Gênes. *Annali della Museo di Storia Naturale in Genova* (2) 19(39): 628–638.

Lesne P. 1899e: Note sur une espèce nouvelle de Bostrychus (Coleoptera: Bostrychidae). *Notes from the Leyden Museum* 20: 255.

Lesne P. 1900a: X. Bostrychidae. P. 149. In: Contribution à l'étude de la faune de Sumatra (Côte ouest – Vice-résidence de Painan). (Chasses de M. J.-L.Weyers). *Annales de la Société Entomologique de Belgique* 44: 147–148.

[1])Lesne P. 1900b: Bostrychides. Pp. 425–427. In: Diagnoses d'insectes du Congo. Insectes du Congo. *Annales de la Société Entomologique de Belgique* 44: 423–435.

[1])Lesne P. 1901a: Révision des Coléoptères de la famille des Bostrychides. 4e Mémoire. Bostrychinae sens. strict. – II. Les Xylopertha. *Annales de la Société Entomologique de France* 69: 473–639.

[1])Lesne P. 1901b: Diagnose d'un type générique nouveau de la tribu des Psoinae [Col.]. *Bulletin de la Société Entomologique de France* 1901: 348–350.

Lesne P. 1901c: Synopsis des Bostrychides paléarctiques. *L'Abeille, Journal d'Entomologie* 30: 73–104.

Lesne P. 1901d: Liste des Bostrychides recueillis en Birmanie par feu M. G.-Q. Corbett. *Annales de la Société de Belgique* 45: 1.

[1)]Lesne P. 1902a: Synopsis des Bostrychides paléarctiques II. *L'Abeille, Journal d'Entomologie* 30: 105–136.

[1)]Lesne P. 1902b: Les Bostrychides indo-chinois du genre Heterarthron [Col.]. *Bulletin de la Société Entomologique de France* 1902: 223–224.

[1)]Lesne P. 1904: Supplément au Synopsis des Bostrychides paléarctiques. *L'Abeille, Journal d'Entomologie* 30: 153–168 + 4 pls.

Lesne P. 1905a: Notes additionnelles et rectificatives sur les Bostrychides paléarctiques. *L'Abeille, Journal d'Entomologie* 30: 249–251.

[1)]Lesne P. 1905b: Note sur un Bostrichide africain (Bostrychopsis villosula nom. nov.). *Bulletin du Muséum d'Histoire Naturelle* 11(5): 298.

[1)]Lesne P. 1905c: Diagnoses de Bostrychides africains nouveaux (Col.). *Bulletin de la Société Entomologique de France* 1905: 275–267.

Lesne P. 1905d: Bostrychides de la Guinée Espagnole. *Memorias de la Real Sociedad Espanola de Historia Natural* 1(11): 197–200.

[1)]Lesne P. 1906a: Bostrychides nouveaux ou peu connus. *Annales de la Société Entomologique de France* 75: 393–428.

[1)]Lesne P. 1906b: Révision des Coléoptères de la famille des Bostrychides. 5e Mémoire. Sinoxylinae. *Annales de la Société Entomologique de France* 75: 445–561.

[1)]Lesne P. 1906c: Note sur une espèce nouvelle des Coléoptères bostrichide, recuelie par M. E.-R. Wagner dans le Chaco argentin. *Bulletin du Muséum National d'Histoire Naturelle* 1906 (1): 12–14.

[1)]Lesne P. 1906d: Note sur deux espèces australiennenes de Bostrichides appartenant au genre Xylobosca. *Bulletin du Muséum National d'Histoire Naturelle* 1906(4): 190–192.

[1)]Lesne P. 1906e: Synopsis des Micrapate de l'Amérique centrale. *L'Abeille, Journal d'Entomologie* 30: 269–281.

[1)]Lesne P. 1906f: Viaggio di Leonardo Fea nell' Africa occidentale. Bostrychidae. *Annali della Museo Civici di Storia Naturali di Genova* (3) 2: 412–417.

Lesne P. 1906g: Nouvelles notes sur les Bostrychides Paléarctiques. *L'Abeille, Journal d'Entomologie* 30: 282.

[1)]Lesne P. 1907a: Diagnose d'un Coléoptère Bostrychidae de l'Amérique du Nord (Scobicia arizonica nov. sp.). *Bulletin du Muséum National d'Histoire Naturelle* 1907(4): 244–245.

Lesne P. 1907b: Mission de M. F. Geay à Madagascar. Diagnose d'un Coléoptère bostrychide du genre Apate (A. Geayi nov. sp.). *Bulletin du Muséum National d'Histoire Naturelle* 1907(18): 324–325.

[1)]Lesne P. 1907c: Un Lyctus african noveau. [Col.]. *Bulletin de la Société Entomologique de France* 1907(5): 302–303.

[1)]Lesne P. 1908a: *7. Coleoptera, 4. Bostrychidae*. Pp. 33–37. Wissenschaftliche Ergebnisse der Schwedischen Zoologischen Expedition dem Kilimadjaro, dem Meru den Umgebenden Massaisteppen Deutsh-Ostafrikas 1905–1906 unter Leitung von Prof. Dr. Yngve Sjöstedt. Dritter Teil, Uppsala: Almquist & Wiksells, 636 pp. + 37 pls.

[1)]Lesne P. 1908b: Notes sur les Coléoptères Térediles. *Bulletin du Muséum National d'Histoire Naturelle* 14(4): 179–181.

[1)]Lesne P. 1908c: Notes sur les Coléoptères Térediles. 2. Un nouveau Lyctidae apparenté au Lyctus brunneus Steph. *Bulletin du Muséum National d'Histoire Naturelle* 14(7): 356–358.

[1)]Lesne P. 1908d: Bostricidae de l'Afrique allemande du Sud-Ouest. In: Schulze L.: Zoologische und Antropologische Ergebnisse einer Forschungreise im westlichen und zentralen Südafrica ausgeführt in den Jahren 1903–1905 mit Unterstützung der Kgl. Preussischen Akademie der Wissenschaften zu Berlin. Erster Band. *Denkschriften der Medizinisch-Naturwissenschaftliche Gesellschaft zu Jena* 13: 425–428.

Lesne P. 1909a: Notes sur le Coléoptères Térediles. 3. Les Lyctides et Bostrychides des archipels Atlantiques. *Bulletin de la Société Entomologique de France* 1909: 347–351.

[1)]Lesne P. 1909b: Révision des Coléoptères de la famille des Bostrychides. 6e Mémoire: Dinapatinae et Apatinae. *Annales de la Société Entomologique de France* 78: 471–574 + 5 pls.

[1)]Lesne P. 1910a: Notes sur divers Lyctides du type Xylotrogus Steph. [Col.]. *Bulletin de la Société Entomologique de France* 1910: 254–255.

[1)]Lesne P. 1910b: Un Lyctide nouveau de la faune indienne [Col.]. *Bulletin de la Société Entomologique de France* 1910: 303–305.

[1)]Lesne P. 1910c: Notes sur le Coléoptères Térediles. 4. Les Bostrychides des Iles Galapagos. *Bulletin du Museum National d'Histoire Naturelle* 1910(4): 183–186.

Lesne P. 1910d: Note III. Liste des Bostrychides et Lycides observés jusqua ce jour dans l'Ille de Java. *Notes from the Leyden Museum* 33: 70–74.

[1)]Lesne P. 1911a: Notes sur les Coléoptères Térediles. 6. Un Lyctidae paléarctique nouveau. *Bulletin du Muséum National d'Histoire Naturelle* 17(2): 48–50.

[1)]Lesne P. 1911b: Notes sur les Coléoptères Térediles. 7. Les Tristariens; leurs affinitiés zoologiques. Synopsis du groupe. *Bulletin du Muséum National d'Histoire Naturelle* 17(4): 202–208.

[1)]Lesne P. 1911c: Notes sur les Coléoptères Térediles. 8. Lyctides nouveaux du Mexique. *Bulletin du Muséum National d'Histoire Naturelle* 17(7): 534–538.

[1)]Lesne P. 1911d: Diagnoses préliminaires de Bostrychides nouveaux du genre Heterarthron [Col.]. *Bulletin de la Société Entomologique de France* 1911: 45–48.

[1)]Lesne P. 1911e: Sur une forme représentative de la faune chilienne en Argentine [Col., Bostrychidae]. *Bulletin de la Société Entomologique de France* 1911: 345–346.

[1)]Lesne P. 1911f: Deux Dinoderus indo-chinois nouveaux [Col., Bostrychidae]. *Bulletin de la Société Entomologique de France* 1911: 397–399.

[1)]Lesne P. 1912a: Diagnose préliminaire d'un nouveau type de Psoien appartennant à la faune indienne (Col., Bostrychidae). *Bulletin de la Société Entomologique de France* 1912: 376–377.

Lesne P. 1912b: Notes sur Notes sur les Coléoptères Térédiles. 10. Les Psoa californiens. *Bulletin de la Société Entomologique de France* 1912: 404–409.

[1)]Lesne P. 1913a: Notes sur les Coléoptères Térédiles. 11. Les Dolichobostrychus et Parabostrychus Indo-malais. *Bulletin du Muséum National d'Histoire Naturelle* 1913 (4): 190–193.

Lesne P. 1913b: Notes sur les Coléoptères Térédiles. 12. – Nouvelles données sur les Psoa de Californie. *Bulletin du Muséum National d'Histoire Naturelle* 19(5): 271–275.

Lesne P. 1913c: Notes sur les Coléoptères Térediles. 13. – Les Tristariens du genre Lyctoderma. *Bulletin du Muséum National d'Histoire Naturelle* 19(8): 562–565.

Lesne P. 1913d: Un Heterarthron argentin nouveau [Col., Bostrichidae]. *Bulletin de la Société Entomologique de France* 1913(8): 191–194.

[1)]Lesne P. 1913e: Missione per la frontiera italo-etiopica sotto il comando del capitano Carlo Citerni. Risultati Zoologici. Liste des Bostrichides et description d'un espèce nouvelle de cette famille. *Annali del Museo Civico di Storia Naturali di Genova* (3) 5 [1911]: 473–475.

[1)]Lesne P. 1914a: Un type nouveau de Dinodériens. Variabilité du tarse chez les Bostrychides [Col.]. *Bulletin de la Société Entomologique de France* 1914: 242–245.

[1)]Lesne P. 1914b: Notes sur les Coléoptères Térediles. 14. Les Lyctides de l'Afrique australe. *Bulletin du Muséum National d'Histoire Naturelle* 20(6): 332–335.

Lesne P. 1914c: Bostrychides et Lyctides (Col.). Pp. 11–14. In: H. Sauter's Formosa – Ausbeute. *Supplementa Entomologica* 3: 1–118.

[1)]Lesne P. 1916: Notes sur les Coléoptères Térediles. 15. Variabilité de certains Lyctides de l'Amerique du Nord. Les formes typiques du genre Lyctus. *Bulletin du Muséum National d'Histoire Naturelle* 22(2): 92–100.

[1)]Lesne P. 1918: Notes sur les Coléoptères Térédiles. 16. Un Sinoxylon indo-malais nouveau (S. parviclava n. sp.). *Bulletin du Muséum National d'Histoire Naturelle* 1918 [24] (7): 490–492.

[1)]Lesne P. 1919: Notes sur les Coléoptères Térédiles. 17. – La série du Sinoxylon capillatum Lsn. Diagnose d'une espéce nouvelle. *Bulletin du Muséum National d'Histoire Naturelle* 25(6): 464–466.

[1)]Lesne P. 1920: Notes sur les Coléoptères Térédiles. 18. Un Bostrychide nouveau de la Faune yunnanaise. *Bulletin du Muséum National d'Histoire Naturelle* 26(4): 295–297.

[1)]Lesne P. 1921a: Le Dinoderus distinctus Lesne et les espèces affines [Col. Bostrychidae]. *Bulletin de la Société Entomologique de France* 1921: 131–134.

[1)]Lesne P. 1921b: Les espèces typiques de Trogoxylon [Col. Bostrychidae]. Position systématique de ce genre. *Bulletin de la Société Entomologique de France* 27(16): 228–231.

[1])Lesne P. 1921c: Classification des Coléoptères Xylophages de la Famille des Bostrychides. *Association Francàise pour L'Avancement des Sciences. Comte Rendu de la 44me Session, Strasbourg 1920*: 285–289.

[1])Lesne P. 1923a: Notes sur les Coléoptères Térédiles. 19. Diagnose préliminaires de Bostrychides nouveaux de l'Afrique tropicale. *Bulletin du Muséum National d'Histoire Naturelle* 29(1): 55–60.

Lesne P. 1923b: Travaux scientifiques de l'Armée d'Orient (1916–1918). Coléoptères: Bostrychidae et Cleridae. *Bulletin du Muséum National d'Histoire Naturelle* 29(3): 240–242.

[1])Lesne P. 1924: *Les Coléoptères Bostrychides de l'Afrique tropicale francàise*. Paris: Les Presses Universitaires de France, Libraire Paul Lechevalier, 301 pp.

Lesne P. 1925a: *Notulae terediliane*. Pp. 25–32 + 1 pl. In: Lesne P.: Encyclopédie Entomologique. Série B. Mémoires et notes I. Coleoptera Etudes sur les Insectes Coléoptères. Tome I. Paris: Paul Lechevalier.

Lesne P. 1925b: *Bostrychidae. In: Angola et Rhodesia (1912–1914). Mision Rohan-Chabot sous les auspices du Ministére de L'instruction publique et de la Société de Géographie. Tome IV Histoire Naturelle. Fascicule 3. Insectes (Coléopteres et Hyménopteres) Arachnides – Mollusques Fougéres*. Paris: P. Geuthner, 246 pp. + Maps + 3 pls.

Lesne P. 1926a: Un Bostrychide chilien peu connu, Neoterius Fairmairei, Lesne. *Revista Chilena de Historia Natural* 30: 23–25.

Lesne P. 1926b: *Fauna Buruana. Coleoptera, Fam. Bostrychidae. In: Boeroe-Expeditie 1921–1922. Résultats zoölogiques de l'Expédition Scientifique Néerlandaise à l'ile de Buru en 1921–1922. Vol II Insecta, Livr. 3.*, Pp. 148–149. Buitenzorg: Archipel.

[1])Lesne P. 1929a: Malayan Bostrychidae (Coleoptera) in the Collection of the F. M. S. Museums. *Journal of the Federated Malay States Museums* 14(3–4): 459–460.

[1])Lesne P. 1929b: Bostrychides recueillis dans la Somalie italienne par la mission Guido Paoli. *Memorie della Società Entomologica Italiana* 8(1): 66–68.

[1])Lesne P. 1930: Diagnoses de Bostrichides nouveaux [Col.]. *Bulletin de la Société Entomologique de France* 1930: 102–104.

[1])Lesne P. 1931a: Notes sur les Coléoptères Térédiles. 20. Diagnoses de Bostrichides nouveau faisant partie des collections du Muséum. *Bulletin du Muséum National d'Histoire Naturelle* (1) 3: 96–105.

Lesne P. 1931b: Un Bostrychidae african nouveau [Col.]. *Bulletin de la Société Entomologique de France* 1931: 24–25.

[1])Lesne P. 1932a: Notes sur les Coléoptères Térediles. 21. Description d'un Sinoxylon nouveau des Iles Philippines. *Bulletin du Muséum National d'Histoire Naturelle* (2) 4: 393–394.

[1])Lesne P. 1932b: Notes sur les Coléoptères Térédiles. 22. Diagnoses de Bostrychides nouveaux de l'Asia orientale. *Bulletin du Muséum National d'Histoire Naturelle* (2) 4: 651–663.

[1)]Lesne P. 1932c: Results of Dr. E. Mjöberg's Swedish Scientific Expeditions to Australia 1910–1913. 52. Bostrychidae. *Arkiv för Zoologi* 24A(14): 1–8.

[1)]Lesne P. 1932d: *Les formes d'adaptation au commensalisme chez les Lyctides*. Pp. 619–627. Paris: Société Entomologique de France, Livre du Centennaire, xii + 729 pp.

[1)]Lesne P. 1933a: Trois Dinoderus indo-malais nouveaux [Col. Bostrychidae]. *Bulletin de la Société Entomologique de France* 38: 257–260.

[1)]Lesne P. 1933b: Sur les Coléoptères Bostrychides du genre Dendrobiella Casey. *Comptes Rendus du Congrès des Sociétés Savantes de Paris et des Département, Section des Sciences* 1933: 237–239.

[1)]Lesne P. 1934a: Sur le genre Hendecatomus Mellié [Col., Bostrychidae]. *Bulletin de la Société Entomologique de France* 39: 174–175.

[1)]Lesne P. 1934b: Sur le Phonapate frontalis Fåhr. [Col., Bostrychidae]. *Bulletin de la Société Entomologique de France* 39: 217–220.

[1)]Lesne P. 1934c: Notes sur les Coléoptères Térédiles. 23e Note – Bostrychides nouveaux des collections du Muséum. *Revue Francqise d'Entomologie* 1(1): 39–43.

[1)]Lesne P. 1934e: Note sur un Bostrichide néo-zélandais l'Euderia squamosa Broun (Coleopt.). *Annales de la Société Entomologique de France* 103(3–4): 389–393.

[1)]Lesne P. 1935a: Bostrychides nouveaux du Congo Belge et des régions voisines. *Revue de Zoologique et Botanique Africaines* 27(1): 1–14.

[1)]Lesne P. 1935b: Un type primitif de Bostrichides le gentre Chilenius. *Annales des Sciences Naturelles, Zoologie* (Série 2), 18(2): 21–38.

[1)]Lesne P. 1935c: *Lyctidae* (*Col.*). Pp. 300–301. In: Visser P. C. & Visser-Hooft J. (eds.): *Wissenschaftliche Ergebnisse niederländischen Expeditionenin den Karakorum un die angrenzenden Gebietein den Jahren 1922, 1925 und 1929/30. Volume 1.* Leipzig: F. A. Brockhaus, xviii + 499 pp. + 1 map + 8 pls.

Lesne P. 1935d: Sur certains rapports faunistiques entre Madagascar et l'Afrique sud-orientale. *Comptes Rendus des Séances de l'Academie de Sciences* 201: 991–992.

Lesne P. 1935e: Quelques précisions sur les Hendecatomus [Col. Bostrychidae]. *Bulletin de la Société Entomologique de France* 40(13–14): 197–199.

[1)]Lesne P. 1935f: Les Bostrichides de l'Arabie. *Revue Francqise d'Entomologie* 1(4): 268–272.

[1)]Lesne P. 1936a: Diagnoses préliminaires de Lyctides et Bostrychides nouveaux [Col.] de l'Indo et l'Austro-Malaisie. *Bulletin de la Société Entomologique de France* 41(9): 131–137.

[1)]Lesne P. 1936b: Contribution à la Faune du Mozambique (Voyage de M. P. Lesne 1928–1929). 22e Note. – Les Enneadesmus du Mozambique et des régions voisines. *Revue Française Entomologique* 3(1): 55–62.

[1)]Lesne P. 1937a: Un genre nouveau de Bostrychides de la région du lac Victoria. *Revue de Zoologiqe et de Botanique Africaines* 29(4): 387–392.

[1])Lesne P. 1937b: Xyloperthini paléarctiques peu connus ou nouveaux (Col. Bostrychidae). *Bulletin de la Société Entomologique de France* 42(13–14): 195–200.

[1])Lesne P. 1937c: Notes rectificatives et synonymiques sur certains Bostrychides [Col.]. *Bulletin de la Société Entomologique de France* 42 (16): 238–240.

[1])Lesne P. 1937d: Une race nouvelle d'Enneadesmus de l'Afrique équatoriale (Coléoptères, Bostrichidae). *Revue de Zoologiqe et de Botanique Africaines* 30(1): 86–87.

[1])Lesne P. 1937e: Un genre peu connu de Bostrychides australiens. *Bulletin de la Société Entomologique de France* 62: 165–174.

[1])Lesne P. 1937f: Bostrychides nouveaux des collections du Muséum (Coléopt.). *Bulletin du Muséum National d'Histoire Naturelle* (Series 2) 9(5): 319–326.

[1])Lesne P. 1938a: Entomological expedition to Abyssinia 1926–27: Coleoptera Bostrychidae. *The Annals et Magazine of Natural History* (Series 2), 10(10): 387–395.

[1])Lesne P. 1938b: Additions à la faune tempérée de la Province du Cap. Les genres Heteropsoa et Dinoderopsis (Col., Bostrichidae). *Bulletin de la Société Entomologique de France* 43(13–14): 170–175.

[1])Lesne P. 1938c: Sur deux Lichenophanes de l'Afrique tropicale (Col. Bostrychidae). *Bulletin de la Société Entomologique de France* 43(15–16): 200–202.

[1])Lesne P. 1938d: Sur un groupe peu connu de Sinoxylon Indo-Malais (Coleoptera, Bostrichidae). *Bulletin de la Société Zoologique de France* 63: 402–415.

Lesne P. 1938e: *Coleopterorum Catalogus. Pars 161 – Bostrychidae*. In: Junk W., Schenkling S. (eds.): Coleopterorum Catalogus. Berlin: W. Junk, 84 pp.

[1])Lesne P. 1939a: Sur le genre Amintinus Lesne (Col. Bostrychidae). *Revue Française Entomologique* 6(2): 33–38.

[1])Lesne P. 1939b: Contribution a l'étude des Bostrychidae de l'Amérique centrale (Coléoptères). *Revue Française Entomologique* 6(3): 91–123.

Lesne P. 1940: Entomological results from the Swedish expedition 1934 to Burma and British India. Coleoptera: Sphaeriidae et Bostrychidae recueillis par René Malaise. *Arkiv för Zoologi* 32(6): 1–4.

[1])Lesne P. 1941a: Sur les genres Xylogenes et Xylomedes. [Col. Bostrychidae]. *Annales de la Société Entomologique de France* 109 [1940]: 131–136.

[1])Lesne P. 1941b: Sur quelques Bostrychides indiens. *Annales de la Société Entomologique de France* 109 [1940]: 137–152.

[1])Lesne P. 1943: Bostrichidae (Coleoptera) Teredilia. Exploration du Parc National Albert. Mission G. F. de Witte (1933–1935). *Institut des Parcs Nationaux du Congo Belge, Bruxelles* 43(7): 29–43.

[3])Letzner K. 1849: [Bostrichus Jalappae]. *Uebersicht der Arbeiten und Veränderungen der Slechischen Gesellschaft für Vaterländischer Kultur* 1848: 99.

[1])Lewis G. 1896: On new Species of Coleoptera from Japan, and Notices of others. *The Annals and Magazine of Natural History* (6) 17: 329–343.

[2)]Lichtenstein A. A. H. 1796: *Catalogus musei zoologici ditissimi Hamburgi, d III Fabruar 1796. Auctionis lege distrahendi. Sectio tertia. Continens Insecta.* 13. Hamburg: G. F. Schniebes, xii + 222 + [2] pp.

[1)]Linell M. L. 1899: On the Coleopterous Insects of the Galapagos Islands. *Proceedings of the United States National Museum* 21(1143) [1888]: 249–268.

[1)]Linnaeus C. 1758: *Systema naturae per regna tria naturae, secundum classes, ordines, genera, species, cum characteribus, differentiis, synonymis, locis. Tomus I. Editio decima, reformata.* Holmiae: Laurentii Salvii, 823 pp.

[1)]Linnaeus C. 1767: *Systema naturae, per regna tria naturae, secundum classes, ordines, genera, species cum characteribus, differentiis, synonimis, locis. Tomus I. Editio decima tertia, ad Editionem duodecimam reformatam Holmiensem.* Vindobonae: I. Thomae de Trattnern 1767: 533–1327 pp. + 37 unpp.

Liu L. Y. 2010: New records of Bostrichidae (Insecta: Coleoptera, Bostrichidae, Bostrichinae, Lyctinae, Polycaoinae, Dinoderinae, Apatinae). *Mitteilungen der Münchner Entomologischen Gesellschaft* 100: 103–117.

[1)]Liu L. Y. 2021a: A review of the powderpost beetle genera Xylothrips Lesne, 1901 and Calophagus Lesne, 1902 (Coleoptera: Bostrichidae: Bostrichinae: Xyloperthini). *European Journal of Taxonomy* 746: 130–147.

Liu L. Y. 2021b: An annotated synopsis of the powder post beetles (Coleoptera: Bostrichidae) of Mainland China. *Zootaxa* 5081(3): 389–419.

[1)]Liu L. Y. & Beaver R. A. 2013a: A new species of Xyloprista Lesne, 1901 with a key of its species (Coleoptera: Bostrichidae: Bostrichinae: Xyloperthini). *Mitteilungen der Münchner Entomologischen Gesellschaft* 103: 95–98.

[1)]Liu L. Y. & Beaver R. A. 2013b: Notes on the genus Prostephanus Lesne, 1897 (Coleoptera: Bostrichidae, Dinoderinae). *Entomologist's Monthly Magazine* 149: 255–262.

[1)]Liu L. Y. & Beaver R. A. 2017: A review of the powderpost beetle genus, Xylopertha Guérin-Méneville, 1845, with a new species and new synonymy (Coleoptera: Bistrichidae: Bostrichinae: Xyloperthini). *European Journal of Taxonomy* 380: 1–22.

Liu L. Y. & Beaver R. A. 2018: A synopsis of the powderpost beetles of the Himalayas with a key to the genera (Insecta: Coleoptera: Bostrichidae), pp. 407–422. In: Hartmann M., Barclay M. & Weipert J. (eds.): *Biodiversität und Naturausstattung im Himalaya VI*. Erfurt: Verein der Freunde und Förderer des Naturkundemuseums Erfurt e.V., 628 pp.

[1)]Liu L. Y. & Beaver R. A. 2023a: The first Asian species of Polycaon Castelnau, and a new montane species of Melalgus Dejean from China (Coleoptera: Bostrichidae: Polycaoninae). *Zootaxa* 5315(3): 271–281.

[1)]Liu L. Y. & Beaver R. A. 2023b: A new West African genus of Bostrichidae (Coleoptera), and a key to the Afrotropical genera of tribe Xyloperthini. *European Journal of Taxonomy* 885: 21–32.

[1)]Liu L. Y., Beaver R. A. & Sanguanshub S. 2016: A new Oriental genus of bostrichid beetle (Coleoptera: Bostrichidae: Xyloperthini), a new synonym and a lektotype designation for Octodesmus episternalis (Lesne, 1901). *European Journal of Taxonomy* 189: 1–12.

[1)]Liu L. Y., Beaver R. A. & Sanguanshub S. 2021: Revision of the genus Xylodrypta Lesne 1901 (Coleoptera: Bostrichidae: Bostrichinae: Xyloperthini) with a new species and a key to species. *Zootaxa* 5005(2): 234–240.

Liu L. Y., Beaver R. A. & Yang J. T. 2006: The Bostrichidae (Coleoptera) of Taiwan: a key to species, new records, and a lectotype designation for Sinoxylon mangiferae Chûjô. *Zootaxa* 1307: 1–33.

Liu L. Y. & Geis K.-U. 2019: A synopsis of the Lyctine beetles of Eurasia with a key to the species (Insecta: Coleoptera: Bostrichidae: Lyctinae). *Journal of Insect Biodiversity* 9: 34–56.

Liu L.-Y., Ghahari H. et Beaver R. A. 2016: An annotated synopsis of the powder post beetles of Iran (Coleoptera: Bostrichoidea: Bostrichidae). *Journal of Insect Biodiversity* 4 (14): 1–22.

[1)]Liu L. Y., Leavengoog J. M. & Bernal M. E. 2022: A preliminary checklist of the Bostrichidae (Coleoptera) of Paraguay. *Boletín del Museo Nacional de Historia Natural del Paraguay* 28(1): 16–36.

[1)]Liu L. Y. & Schönitzer K. 2011: Phylogenetic analysis of the family Bostrichidae auct. at suprageneric levels (Coleoptera: Bostrichidae). *Mitteilungen der Müncher Entomologischen Gesellschaft* 101: 99–132.

Liu L. Y., Schönitzer K. & Yang J. T. 2008: A review of the literature on the life history of Bostrichidae. *Mitteilungen der Münchner Entomologischen Gesellschaft* 98: 91–97.

Liu L. Y., Schönitzer K. & Yang J. T. 2009: Microstructural characters as a tool for taxonomy (Coleoptera: Bostrichidae: Minthea and Dinoderus). *Nachrichtenblatt der Bayerischen Entomologen* 58 (3–4): 58–61.

[1)]Liu L. Y. & Sittichaya W. 2022: The Oriental of Xyloperthini (Coleoptera: Bostrichidae: Bostrichinae) with a new genus and species from Thailand, and a key to the genera. *European Journal of Taxonomy* 828: 45–60.

Löbl I. & Smetana A. 2007: *Catalogue of Palaearctic Coleoptera. Elateroidea – Derodontoidea – Bostrichoidea – Lymexyloidea – Cleroidea – Cucujoidea. Volume 4.* Stenstrup: Apollo Books, 935 pp.

López-Colón J I. 2000: Los Bostrichidae Lateille, 1802 de la fauna de Marruecos (Coleoptera). *Biocosme Mésogéen, Nice* 16(4) [1999]: 171–221.

[1)]Lucas P. H. 1843a: Sur plusieurs Coléoptères nouveaux du nord de l'Afrique. *Revue Zoologique* 6: 158–159.

[1)]Lucas P. H. 1843b: Sur plusieurs Coléoptères nouveaux du nord de l'Afrique nés à Paris. *Bulletin de la Société Entomologique de France* 1843: 24–26.

[1)]Lucas P. H. 1849: *Exploration scientifique de l'Algérie pendant les années 1840, 1841, 1842.publiée par ordre du gouvenrnement et aves le concours d'une Commission Académique. Sciencies physiques, Zoologie. Vol. II. Histoire naturelle des animaux articulés. Cinquième classe. Insectes. Premier Ordre. Les coléoptères.* Paris: Imprimerie Nationale [1849], 590 pp. + 47 pl. [Pp. 449–590 issued 1849].

[1)]Lucas P. H. 1850: Observations sur un nouveau genre de l'ordre des Coléoptères (Genus Stenomera) qui habite les possessions françaises du nord de l'Afrique. *Annales de la Société Entomologique de France* (2) 8: 31–43 + 1 pl.

[1)]MacLeay W. J. 1873: Notes on a collection of Insects from Gayndah. *The Transactions of the Entomological Society of New South Wales* 2: 239–318.

[1)]MacLeay W. J. 1886: The insects of the Fly River, New Guinea, "Coleoptera". *Proceedings of the Linnaean Society of New South Wales* (2) 1: 136–157.

Majka C. G. 2007: The Derodontidae, Dermestidae, Bostrichidae, and Anobiidae of the Maritime Provinces of Canada (Coleoptera: Bostrichiformia). *Zootaxa* 1573: 1–38.

[3)]Mannerheim C. G. von 1843: Beitrag zur Kaefer-fauna der Aleutischen Inseln, der Insel Sitkha und Neu-Californiens. *Bulletin de la Société Impériale des Naturalistes de Moscou* 16: 175–314.

[3)]Mannerheim C. G. von 1852: Zweiter Nachtrag zur Kaefer-fauna der Amerikanischen Laender des Russischen Reiches. *Bulletin de la Société Impériale des Naturalistes de Moscou* 25(2): 283–387.

[1)]Marseul S. A. de 1857: *Catalogue des coléoptères d'Europe*. Paris, xvi + 200 pp.

[1)]Marseul S. A. de 1867: Descriptions d'espéces nouveles. *L'Abeille, Mémoires d'Entomologie* 4: 33–40.

[1)]Marseul S. A. de 1879: Mélanges (suite). In: Nouvelles et faits divers de L'Abeille. Deuxieme Série – No 21 [1878]. *L'Abeille, Journal d'Entomologie* (Série 3) 5 (21): 82–84.

[1)]Marseul S. A. de 1883: Mélanges (suite). In: Nouvelles et faits divers de L'Abeille. Deuxieme Série, 46. *L'Abeille, Journal d'Entomologie* (Série 4) 3: 182–184.

[1)]Marsham T. 1802: *Entomologia Britannica, sistens insecta britanniae indigena, secundum methodum linnaeanam disposita. Tomus I. Coleoptera.* Londini: Wilks et Taylor, J. White, xxxi + 548 pp.

Mateu J. 1965: Notes sur quelques Cérambycides, Lyctides et Bostrychides de la région de Béni-Abbés (Sahara Nord-occidental). *L'Entomologiste* 21: 103–114.

[1)]Mateu J. 1974: Les Bostrychides des environs de Makokou et de Belinga (Rép. Du Gabon). *Nouvelle Revue d'Entomologie* 4(1): 39–49.

Mathur R. N. 1955: Identification of Indian Lyctidae and Dinoderus species. *Indian Forest Bulletin, N. S., Entomology* 187: 1–8 + 1 pl.

[1)]Matsumura S. 1911: Erster Beitrag zur Insekten-Fauna von Sachalin. *The Journal of the College of Agriculture, Tohoku Imperial University* 4 (1): 1–145 + 2 pl.

[1)]Matsumura S. 1915a: *Konchu Bunruigaku II.* [Insect Taxonomy]. Tokyo: Keiseisha, 316 + 10 + 10 pp.

Matsumura S. 1915b: *Dai Nippon Gaichu Zensho*. Tokyo: Rokumeikan, 308 + 96 pp.

[1]Mellié J. 1847: Mélanges et nouvelles. *Revue Zoologique* 10: 108–110.

[1]Mellié J. 1848: Monographie de l'ancien genre Cis des auteurs. *Annales de la Société Entomologique de France* (2) 6: 205–274.

[1]Melsheimer F. E. 1806: *A Catalogue of Insects of Pennsylvania. Part First*. Hanover, York County: W. D. Lepper, vi + 60 pp.

Melsheimer F. E. 1845: *Catalogue of the described Coleoptera of the United States*. Washington: Smithsonian Institute, xvi + 174 pp.

[1, 3]Melsheimer F. E. 1846: Descriptions of new species of Coleoptera of the United States. *Proceedings of the Academy of Natural Science of Philadelphia* 2: 98–118.

[1]Mendes L. O. T. 1932: Una nova especie de genero Neoterius (Col., Bostrychidae) broca de Vitis vinifera L. *Revista de Entomologia* 2(1): 27–32.

Meurgey F. & Ramage T. 2020: Challenging the Wallacean shortfall: A total assessment of insect diversity on Guadeloupe (French West Indies), a checklist and bibliography. *Insecta Mundi* 0786: 1–183.

Mifsud D. & Nardi G. 2016: First Maltese record of Stephanopachys quadricollis (Marseum, 1879) (Coleoptera, Bostrichidae). *ZooKeys* 606: 65–75.

[1]Montrouzier X. 1857: *Essai sur la Faune de l'Ille de Woodlark ou Moiou*. Lyon: F. Dumoulin, 1–114 pp.

[1, 3]Montrouzier X. 1861: Essai sur la faune entomologique de la Nouvelle-Calédonie (Balade) et des iles des Pins, Art, Lifu, etc. Coléoptères (Fin) (1). *Annales de la Société Entomologique de France* (4) 1: 265–306.

Morrone J. J. 2001: *Biogeografía de América Latina y el Caribe*. Zaragoza: M&T-Manuales & Tesis SEA, 3, 148 pp.

[1]Motschulsky V. I. von 1845: Remarques sur la collection de Coléoptères Russes. 1er Article. *Bulletin de la Société Impériale des Naturalistes de Moscou* 18: 1–127 + 3 pl.

[1]Motschulsky V. I. von 1853: Notices. *Etudes Entomologiques* 1: 21–24.

Müller O. F. 1764: *Fauna insectorum Fridrichsdalina, sive methodica descriptio insectorum agri fridrichsdalensis cum characteribus genericis et specificis, nominibus trivialibus, locis natulibus, iconibus allegatis, novisque pluribus speciebus additis*. Hafniae et Lipsiae: F. Gleditschi, xxiv + 96 pp.

[3]Müller P. W. J. 1807: Avis sur une espèce de Bostriche qui détruit les racines du trèfle des prés. *Mémoires de la Société des Artes et Sciences du Département du Mont-Tonerre* 1: 47–64.

[3]Müller P. W. J. 1818: Vermischte Bemerkungen über einige Käferfarten. *Magazin der Entomologie* 3: 228–260.

[3]Müller P. W. J. 1821: Neue Insecten, beschreiben von Müller. *Magazin der Entomologie* 4: 184–230.

[1)]Mulsant E. 1851: Note sur le Bostrichus trispinosus d'Olivier. *Mémoires de l'Académie Nationale des Sciences, Belles-Lettres et Arts de Lyon, Classe des Sciences* (N.S.) (2) 1: 205–208.

[3)]Mulsant E. & Rey C. 1853: Description d'une espèce nouvelle de Coléoptère du genre Bostrichus. *Opuscules Entomologiques de Lyon* 2: 91–92.

[3)]Mulsant E. & Rey C. 1857a: Description d'une nouvelle espèce de coléoptère du genre Bostrichus. *Opuscules Entomologiques* 7 [1856]: 111–113. [30-V-1857]

[3)]Mulsant E. & Rey C. 1857b: Description d'une nouvelle espèce de Coléoptère du genre Bostrichus. *Annales de la Société Linnéenne de Lyon* (N.S.) 3 [1856]: 111–113. [5-IX-1857]

[1)]Mulsant E. & Wachanru 1852: Première série de Coléoptère nouveaux ou peu connus. *Mémoires de la Acedemie Scientes de Lyon* (2) 2: 1–17.

[1)]Murray A. 1867: List of Coleoptera received from Old Calabar, on the West Coast Africa. *The Annals and Magazine of Natural History* (3) 20: 83–95.

Nardi G. & Mifsud D. 2015: The Bostrichidae of the Maltese Islands (Coleoptera). *ZooKeys* 481: 69–108.

Németh T., Kotán A. & Merkl O. 2015: First record of Apate monachus in Hungary, with a checklist of and a key to the Hungarian powderpost beetles (Coleoptera: Bostrichidae). *Folia Entomologica Hungarica* 76: 99–105.

Nong L.-V. 1973: Contributions to the knowledge of the family Bostrychidae (Coleoptera) from North Vietnam. *Travaux du Muséum d'histoire Naturelle „Grigore Antipa"* 13: 155–172.

[3)]Nördlinger H. 1848: Nachtrag zu Ratzeburg Forstinsecten. *Entomologische Zeitung* 9 (8): 225–271.

[1)]Nördlinger H. 1855: *Die kleinen Feinde der Landwirthschaft oder Abhalun der in Feld, Garten und Haus schädlichen oder lästigen Kerfe, sonstigen Gliederthierchen, Würmer, und Schnecken, mit besonderer Berücksichtigung ihrer natürlichen Feinde und der gegen sie anwendbaren Schutzmettel.* Stuttgart und Augsburg: J. G. Cotta, xxiv + 637 pp.

[3)]Nördlinger H. 1856: *Nachträge zu Ratzeburg Forstinsekten. Progreamm.* Stuttgart: Weisse 83 pp.

[1)]Nördlinger H. 1880: *Lebensweise von Forstkerfen oder Nachträge zu Ratzeburg's Forstinsekten, zweite vermehrte Auflage.* Stuttgart: J. C. Cotta v + 73 pp.

[1)]Olivier A. G. 1790a: *Encyclopédie méthodique, ou par ordre de matières; par une société de gens de lettres, de savans et d'artistes; précédée d'un vocabulaire universel, servant de table pour tout l'ouvrage, ornée des portraits de Mm. Diderot et d'Alembert, premiers éditeurs de l'Encyclopédie. Histoire Naturelle. Insectes. Tome sixieme.* Paris: Panckoucke, 793 pp.

[1), 3)]Olivier A. G. 1790b: *Entomologie, ou histoire naturelle des Insectes, aves leurs caractères génériques et spécifiques, leur description, leur synonymie, et leur figure enluminée. Coléoptères. Tome quatrième.* Paris: de Lanneau, No 77, 1–18, Plates 1–2.

OPINION 1754 (1994a): Histoire abrégée des insectes qui se trouvent aux environs de Paris (Geoffroy, 1762): some generic names conserved (Crustacea, Insecta). *Bulletin of Zoological Nomenclature* 51(1): 58–70.

OPINION 1771 (1994b): Cryptophagus advena Waltl, 1834 (currently Ahasverus advena; Insecta, Coleoptera): specific name conserved. *Bulletin of Zoological Nomenclature* 51: 172–173.

OPINION 1811 (1995): COLYDIIDAE Erichson, 1842 (Insecta, Coleoptera): given precedence over CERYLONIDAE Billberg, 1820 and ORTHOCERINI Blanchard, 1845 (1820); and Cerylon Latreille, 1802: Lyctus histeroides Fabricius, 1792 designated as the type species. *Bulletin of Zoological Nomenclature* 52(2): 214–216.

Oromí P., Martín E., Zurita N. & Cabrera A. 2005: Coleoptera. Pp. 78–86. In: Arechavaleta M., Zurita N., Marrero J. L. & Martín J. L. (eds.): Lista preliminar de especies silvestres de Cabo Verde (hongos, plantas, y animales terrestres). *Consejería de Medio Ambiente y Ordenación Territorial, Gobierno de Canarias*, 155 pp.

Özgen I. & Háva J. 2021: Xylopertha reflexicauda (Lesne, 1937) a new species to the Turkish fauna with a new pest record on fig in Turkey. *Entomological News* 130(1): 113–116.

[1)]Pallas P. S. 1772: *Spicilegia Zoologica qiubus novae imprimis et obscurae animalium species iconibus, decriptionibus atque commentariis illustrantur. Fasciculus Nonus.* Berolini: G. A. Lange, 86 pp.

[3)]Panzer G. W. F. 1791: Beschreibung eines noch unbekannten sehr kleinen Kaputkäfer aus einem westindischen Saamen. *Naturforscher* [Halle] 25: 35–38.

[1), 3)]Panzer G. W. F. 1793a: *Faunae insectorum Germanicae initia oder Deutschlands Insecten. Erster Iahrgang, Heft 4.* Nürnberg: Felsecker, 24 sheets + 24 pls.

[3)]Panzer G. W. F. 1793b: *Faunae insectorum Germanicae initia oder Deutschlands Insecten. Erster Iahrgang, Heft 8.* Nürnberg: Felsecker, 24 sheets + 24 pls.

[1), 3)]Panzer G. W. F. 1794a: *Faunae insectorum Germanicae initia oder Deutschlands Insecten. Heft 15.* Nürnberg: Felsecker, 24 sheets + 24 pls.

Panzer G. W. F. 1794b: *Faunae insectorum Germanicae initia oder Deutschlands Insecten. Zweyter Iahrgang, Heft 24.* Nürnberg: Felsecker, 24 sheets + 24 pls.

[3)]Panzer G. W. F. 1795: *Deutschlands Insectenfaune oder entomologisches Taschenbuch für das Iahr 1795.* Nürnberg: Felssecker, [32] + 370 + [2] pp. + 12 pls.

[3)]Panzer G. W. F. 1796: *Faunae insectorum Germanicae initia oder Deutschlands Insecten. Heft 35.* Nürnberg: Felsecker, 24 sheets + 24 pls.

[2)]Panzer G. W. F. 1797: *Johann Euseb Voets Beschreibungen und Abbildungen hartschaaligter Insecten. Coleoptera Linn. Aus dem Original getreu übersetzt mit der in selbigem fehlenden Synonymie und beständigen Commentar versehen vov Dr. Georg Wolfgang Franz Panzer. Teil 4.* Nürnberg: Valentin Bischoff, 14 + 120 pp. + pls 25–48.

[3)]Panzer G. W. F. 1799: *Faunae insectorum Germanicae initia oder Deutschlands Insecten. Heft 66.* Nürnberg: Felsecker, 24 sheets + 24 pls.

[3]Panzer G. W. F. 1805: *Faunae insectorum Germanicae initia oder Deutschlands Insecten. Heft 98*. Nürnberg: Felsecker, 24 sheets + 24 pls.

[1]Panzer G. W. F. 1807: *Faunae insectorum Germanicae initia oder Deutschlands Insecten. Heft 101*. Nürnberg: Felsecker, 24 sheets + 24 pls.

Park S., Jung J.-K. & Han Y.-E. 2020: New Record of Two Species in the Family Bostrichidae (Coileoptera) to Korean Fauna. *Entomological Research Bulletin* 36: 19–21.

[1]Park S., Lee S. & Hong K.-J. 2015: Review of the family Bostrichidae (Coleoptera) of Korea. *Journal of Asia-Pacific Biodiversity* 8: 298–304.

Parnaudeau R. 2017: Contribution à l'inventaire de l'entomofaune de Mayotte. Liste des coléoptères. *Contribution à l'étude des Coléoptères de La Réunion et des archipels de l'océan Indien occidental. Tome II. Supplément au Bulletin de liaison d'ACOREP-France « Le Coléoptériste »*: 48–57.

[1]Pascoe F. P. 1866a: List of the Colydiidae collected in the Amazons Valley by H. W. Bates, Esq., and Descriptions of a new Species. *The Journal of Entomology* 2: 79–99 + pl. 5.

[1]Pascoe F. P. 1866b: List of the Colydiidae collected in the Indian Islands by Alfred R. Wallace, Esq., and Descriptions of New Species. *The Journal of Entomology* 2: 121–143 + pl. 8.

[1], [3]Paykull G. F. von 1800: *Fauna suecica. Insecta. Tomus III*. Upsaliae: J. F. Edman, 459 pp.

Peck S. B. 2005: *A checklist of the beetles of Cuba with data on distributions and bionomics (Insecta: Coleoptera). Arthropods of Florida and Neighboring land Areas. Volume 18*. Gainesville, 249 pp.

Peck S. B. 2006: The beetle fauna of Dominica, Lesser Antilles (Insecta: Coleoptera): Diversity and distribution. *Insecta Mundi* 20(3–4): 165–209.

Peck S. B. 2009a: The beetles of Barbados, West Indies (Insecta: Coleoptera): diversity, distribution and faunal structure. *Insecta Mundi* 73: 1–51.

Peck S. B. 2009b: The beetles of St. Lucia, lesser Antilles (Insecta: Coleoptera): diversity and distributions. *Insecta Mundi* 106: 1–34.

Peck S. B. 2010: The beetles of the island of St. Vincent, Lesser Antilles (Insecta: Coleoptera); diversity and distributions. *Insecta Mundi* 144: 1–77.

Peck S. B. 2011: The beetles of Martinique, Lesser Antilles (Insecta: Coleoptera); diversity and distributions. *Insecta Mundi* 178: 1–57.

Peck S. B. 2016: The beetles of the Lesser Antilles (Insecta, Coleoptera): diversity and distributions. *Insecta Mundi* 460: 1–360.

Peck S. B., Heraty J., Landry B. & Sinclair B. J. 1998: The introduced insect fauna of an oceanic archipelago: The Galapagos Islands, Ecuador. *American Entomologist* 44: 218–237.

Peck S. B. & Thomas M. C. 1998: *A distributional checklist of the beetles (Coleoptera) of Florida, Arthropods of Florida and Neighboring Land Areas*. Florida Department of Agriculture and Consumer Services 16, viii + 180 pp.

Peck S. B., Thomas M. C. & Turnbow R. H. 2014: The diversity and distributions of the beetles (Insecta: Coleoptera) of the Guadeloupe Archipelago (Grande-Terre, Basse-Terre, La Désirade, Maria-Galante, Les Saintes, and Petite-Terre), Lesser Antilles. *Insecta Mundi* 352: 1–156.

1)Peng Y., Jiang R., Shi C., Song W., Long X., Engel M. S. & Wang S. 2022: Alitrepaninae, a new subfamily of auger beetles from mid-Cretaceous Kachin amber of northern Myanmar (Coleoptera: Bostrichidae). *Cretaceous Research* 137(105244): 1–6.

1)Perkins R. C. L. & Sharp D. 1910: Family: Bostrichidae. Pp. 642–644. In: Perkins R. C. L., Scott H. & Sharp D.: *Fauna Hawaiiensis of the Zoology of the Sandwich (Hawaiian) Isles. Coleoptera. Volume III. Part VI. Coleoptera. IV*. Cambridge: The University Press, 666 pp.

1)Peris D., Delclos X., Soriano C. & Perrichot V. 2014: The earliest occurrence and remarkable stasis of the family Bostrichidae (Coleoptera: Polyphaga) in Cretaceous Charentes amber. *Palaeontologica Electronica* 17(4): 1–8.

3)Perroud B.-P. & Montrousier P. 1864: Essai sur la faune entomologique de Kanala (Nouvelle-Calédonie) et description de quelques espèces nouvelles ou peu connues. *Annales de la Société Linnéenne de Lyon (Nouvele Série)* 11: 46–257.

1)Perty J. A. M. 1832: *Delectus animalium articulatorum, quae in itinere per Brasiliam annis MDCCCXVII–MDCCCXX jussu et auspiciis Maximiliani Josephi I. Bavariae Regis Augustissimi peracto collegerunt Dr. J. B. de Spix et Dr. C. F. Ph. de Martius. Digessit, descriptis, pingenda curavit Dr. Maximilianus Perty, praefatus est et edidit Dr. C. F. Ph. De Martius. Fasc. 2*. München: Perty, 61–124 pp. + pls. 13–24.

Pfeffer A. 1940: Notulae Ipidologicae II. Symbolae ad cognitionem generis Pityophthorus Eichh. *Sborník Entomologického Oddělění Národního Muzea v Praze* 18(182): 107–127.

2)Philippi R. A. & Philippi F. H. E. 1864: Beschreibung einiger neuen Chilenischen Käfer. *Entomologische Zeittung* 25(10–12): 313–406.

Pic M. 1893: Notes sur quelques Coléoptères, avec descriptions. *L'Échange, Revue Linnéenne* 9(107): 122.

1)Poinar G., jr. 2013: Discoclavata dominicana n. gen., n. sp., (Coleoptera: Bostrichidae) and Lissantauga epicrana n. ge., n. sp., (Coleoptera: Eucnemidae) in Dominican amber. *Historical Biology* 25(1): 107–113.

Ponel P., Fadda S., Lemaire J.-M., Matocq A., Cornet M. & Pavon D. 2011: *Arthropodes de la principauté de Monaco. Coléoptères, Hétéroptères. Aperçu sur les Fourmis, les Isopodes et les Pseudoscorpions. Monacobiodiv. Rapport final – 1er février 2011*. Principauté de Monaco. Direction de l'Environnement, 100 pp.

[1])Portevin G. 1931: *Histoire naturelle des Coléopteres de France. Tome II. Polyphaga: Lamellicornia, Palpicornia Diversicornia. Encyclopédie Entomologique. Série A [Tome XIII]*. Paris: Paul Lechevalier & Fils, vi + [1] + 542 pp. + pls. 6–10.

[1])Quedenfeldt F. O. G. 1887: Verzeichniss der von Herrn Major a. D. von Mechow in Angola und am Quango-Strom 1878–1881 gesammelten Anthotribiden und Bostrychiden. *Berliner Entomologische Zeitschrift* 30 [1886]: 303–328 + pl. 8.

[1])Rai K. 1965: Studies on the morphology and taxonomy of Indian Bostrychidae VI. A new species of the genus Bostrychopsis Lesne from India (Coleoptera, Bostrychidae). *Journal of the Bombay Natural History Society* 62(1): 576–578.

Rai K. 1966: Studies on the morphology and taxonomy of Indian Bostrychidae. VII. – A revision of the Indian species of Bostrychopsis Lesne (Coleoptera, Bostrychidae). *Bulletin of Entomology. Department of Zoology, Loyola College* 1966(7): 20–25.

[1])Rai K. 1967a: Studies on the morphology and taxonomy of Indian Bostrychidae VIII. – A new species of the genus Xyloscopus Lesne from India (Coleoptera, Bostrychidae). *Science and Culture* 33(3): 140–141.

Rai K. 1967b: Studies on the morphology and taxonomy of Indian Bostrychidae X. A revision of the Indian species of *Melalgus* Dejean, 1835. *Oriental Insects* 1(1–2): 21–27.

Rai K. 1971: Studies on the morphology and taxonomy of Indian Bostrychidae (Coleoptera). XI. A revision of the Indian species of Octodesmus Lesne. *Eos, Revista Española de Entomologia* 46: 351–357.

[1])Rai K. 1978: Studies on the taxonomy and morphology of Indian Bostrychidae XII. A new species of Xyloprista Lesne from India (Coleoptera: Bostrychidae). *Oriental Insects* 12(1): 119–121.

[1])Rai K. & Chatterjee P. N. 1963a: Studies on the morphology and taxonomy of Indian Bostrychidae III. – A new species of Sinoxylon Duftschm. from North India (Coleoptera: Bostrychidae). *Journal of the Timber Dryer's and Preservers Association of India* 9(4): 15–17.

Rai K. & Chatterjee P. N. 1963b: Studies on the morphology and taxonomy of Indian Bostrychidae V. – A revision of the Indian species of Heterobostrychus Lesne (Coleoptera: Bostrychidae). *Indian Forest Records* (N.S.) 10: 205–218.

[1])Rai K. & Chatterjee P. N. 1964: Studies on the morphology and taxonomy of Indian Bostrychidae IV. A new species of Xylopertha Guerin-Meneville (Coleoptera: Bostrychidae) from North India. *The Indian Forester* 90(2): 122–124.

[1])Randall J. W. 1838: Descriptions of new species of Coleopterous Insects inhabiting the State of Maine. *Boston Journal of the Natural History* 2: 1–33.

[3])Ratzeburg J. T. C. 1837: *Die Forst-Insecten oder Abbildung und Beschreibung der in den Wäldern Preussens und der Nachbarstaaten als schädlich oder nützlich bekannt gewordenen Insecten. Erster Theil. Die Käfer*. Berlin: Nicolai, x + 4 + 202 pp. + xxi pls.

[1])Redtenbacher L. 1847: *Fauna Austriaca. Die Käfer. Nach der analytischen Methode*. Heft 3. Wier: K. Gerold's Sohn, 321–480 pp.

[1)]Redtenbacher L. 1868: *Reise der Österreichischen Fregatte Novara um die Erde in den Jahren 1857, 1858, 1859 unter den Befehlen des Commodore B. von Wüllerstorf-Urbair Zoologischer Theil. Zweiter Band. Coleopteren. Mit Fünf Tafeln.* Wien: K. Gerold's Sohn, iv + 249 + 5 pp. + 5 pls.

[1)]Reichardt H. 1962a: Bostrychidae (Coleoptera) 1: Notas sôbre Bostrychopsis neotropicais com descrição de uma nova espécie. *Papéis Avulsos do Departemento de Zoologia Secretaria da Agricultura* 15(2): 15–21.

Reichardt H. 1962b: Bostrychidae (Coleoptera) 2: Dolichobostrychus vitis (Mendes, 1932), sinônimo de Dolichobostrychus angustus (Stein., 1874). *Papéis Avulsos do Departemento de Zoologia Secretaria da Agricultura* 15(6): 63–65.

[1)]Reichardt H. 1962c: Bostrychidae (Coleoptera) 3: Novo gênero e espécie da subtribu Xyloperthina. *Papéis Avulsos do Departemento de Zoologia Secretaria da Agricultura* 15(15): 173–176.

Reichardt H. 1964a: Bostrychidae (Coleoptera) 4. On the Bostrychidae in Blackwelder's "checklist". *Papéis Avulsos do Departemento de Zoológico Secretaria da Agricultura – Sao Paolo* 16: 71–72.

Reichardt H. 1964b: Bostrychidae (Coleoptera) 5. Sobre a distribuciao geografica de duas especies de Apate introduzidas no Brazil. *Papéis Avulsos do Departemento de Zoológico Secretaria da Agricultura – Sao Paolo* 16: 105–108.

Reichardt H. 1964c: Bostrichidae (Coleoptera). 6. Notas sôbre a distribuição geográfica de espécies americanas. *Revista Brasileira de Entomologia* 11: 37–45.

[1)]Reichardt H. 1966: Bostrychidae (Coleoptera) 7. A new Xylothrips from China. *The Coleopterists' Bulletin* 20(3): 81–83.

[1)]Reichardt H. 1970: XIV – Bostrychidae (Coleoptera) from the Galapagos Islands (1) – In: Resultates scientifiques du Mission zoologique belge aux îles Galapagos et en Equador (N. et J. Leleup 1964–1965). Résultates scientifiques. Deuxieme Partie. *Museum Royale de l'Afrique Centrale* 1970: 213–223.

[1)]Reitter E. 1878: Beiträge zur Kenntniss aussereuropäischer Coleopteren. *Entomologische Zeitung* [Stettin] 39(7–9): 314–322.

[1)]Reitter E. 1879: Beitrag zur Kenntniss der Lyctidae. *Verhandlungen der Kaiserlich-königlichen Zoologisch-botanischen Gesellschaft in Wien* 28 [1878]: 195–199.

[1)]Reitter E. 1889: Coleopterologische Ergebnisse der im Jahre 1886 und 1887 in Transcaspien von Dr. G. Radde, Dr. A. Walter und A. Konchin ausgeführten Expedition. *Verhandlungen des Naturforschenden Vereines in Brünn* 27 [1888]: 95–133.

[1)]Reitter E. 1898: Eine Decade neuer Coleopteren aus der Buchara. *Wiener Entomologische Zeitung* 17: 10–16.

[2)]Richter C. F. W. 1820: *Suplementa Faunae Insectorum Europae. Fasciculus I.* Vratislaviae: R. F: Schoene, [3] + 12 pp. + 12 pls.

[1)]Roberts H. 1967: A new powder-post beetle, Xyloperthella guineensis, together with an annotated Check List of the subfamily Bostrychinae (Col., fam. Bostrychidae) from Nigeria. *Journal of Natural History* 2(1) [sic! 1968]: 85–104.

[3]Rosenhauer W. G. 1856: *Die Thiere Andalusiens nach dem Resultate einer Reise zusammengestellt, nebst den Beschreibungen von 249 neuen oder bis jetzt noch unbeschrieben Gattungen und Arten.* Erlangen: T. Blaesing, viii + 429 pp. + 3 pls.

[1], [2]Rossi R. 1790: *Fauna Etrusca, sistens Insecta quae in provinciis Florentina et Pisana praesertim collegit. Tomus primus.* Liburni: T. Masi & Sociorum, xxii + 272 pp.

[1]Rossi R. 1792: *Mantisa Insectorum exhibens species nuper in Etruria collectas a Petro Rossio adiectis faunae Etruscae illustrationibus, ac emendationibus. 1.* Pisa: Polloni, 148 pp.

[3]Sahlberg C. R. 1836a: *Dissertatio entomologica Insecta Fenica enumerans. 2. Part 9. Respondente J. J. Staudinger.* Helsingforsiae, pp. 129–144.

[3]Sahlberg C. R. 1836b: *Dissertatio entomologica Insecta Fenica enumerans. 2. Part 10. Respondente A. Wacklin.* Helsingforsiae, pp. 145–160.

[1]Santoro F. H. 1956: Una nueva especia de la subfamilia Trogoxyloninae (Col. Lyctidae). *Revista de la Sociedad Entomológica Argentina* 18(3–4) [1955]: 45–48.

[1]Santoro F. H. 1957a: Trogoxylon giacobbii, nueva especie Argentina (Coleoptera, Lyctidae). *Revista de Investigaciones Forestales* 1(1–2): 153–155.

Santoro F. H. 1957b: Nuevo lictido (Coleopt.) para la fauna argentina Phyllyctus gounellei (Grouv.) Lesne. Redescription y primeros datos biologicos. *Revista de la Sociedad Entomológica Argentína* 19: 69–71.

Santoro F. H. 1959: Los líctidos (Col.) de la colección del Museo Argentino de Ciencias Naturales Bernardino Rivadaria. *Revista de la Sociedad Entomológica Argentina* 21(3–4) [1958]: 97–98.

[1]Santoro F. H. 1960a: Descripcion de un Lictido (Col.) nuevo de la Argentina. *Revista de Investigaciones Forestales* 2(1): 101–102 + 1 pl.

[1]Santoro F. H. 1960b: Nueva especie Argentina del género Lyctus (Coleoptera-Lyctidae). Pp. 187–190. In: *Acta Trabajos del Primer Congreso Sudamericano de Zoologia (La Plata 18–23 Octobre 1959). Tomo III. secció IV. Entomologia.* La Plata, 276 pp.

[1]Santoro F. H. 1960c: Los Lictidos (Col.) de Museo de La Plata y descripción de does especies nuevas. Pp. 191–197. In: *Acta Trabajos del Primer Congreso Sudamericano de Zoologia (La Plata 18–23 Octobre 1959). Tomo III. secció IV. Entomologia.* La Plata, 276 pp.

Santoro F. H. 1965: Redescription de Trogoxylon recticolle Reitter (Coleoptera: Lyctidae). *Revista de la Sociedad Entomológica Argentina* 27(1–4) [1964]: 11–13.

Santoro F. H. & Viedma M. G. de 1965: Los Líctidos del Instituto Español de Entomología (Coleoptera). *Graellsia* 21: 85–87.

[1]Say T. 1824: Descriptions of Coleopterous Insects collected in the late Expedition to the Rocky Mountains, performed by order of Mr. Calhoun, Secretary of War, under the command of Major Long. Continuation. *Journal of the Academy of Natural Science of Philadelphia* 3: 298–331.

[1], [3]Say T. 1826: Descriptions of new species of Coleopterous Insects inhabiting the United States. *Journal of the Academy of Natural Science of Philadelphia* 5: 237–284.

Schäfer K., Goergen G. & Borgemeistaer Ch. 2000: An illustrated identification key to four different species of adult Dinoderus (Coleoptera: Bostrichidae), commonly attacking dried cassava chips in West Africa. *Journal of Stored Products Research* 36: 245–252.

[1]Schaufuss L. W. 1882: Zoologische Ergebnisse von Excursion auf den Balearen. III. Adenda und Forsetzung, Nunquam Otiosusd. *Zoologische Mittheilungen* 3: 527–552.

[1]Schedl K. E. 1964: Forstentomologische Beiträge aus Madagaskar Familie Bostrychidae. *Journal of Applied Entomology* 55(1–4): 276–287.

Schedl K. E. 1972: Contribution à l'Entomologie forestière du Congo. *Publications de l'Institut National pour l'Étude Agronomique du Congo* (*I.N.É.A.C.*), No.114: 1–68.

[1]Schilsky J. 1899: *Die Käfer Europa's. Nach der Natur beschreiben von Dr. H. C. Küster und Dr. G. Kraatz. Fortgesetzt von J. Schilsky. Heft 36*. Nürnberg: von Bauer und Raspe. iv + 100 pls. [354] pp.

[1]Schönfeldt H von 1887: Catalog der Coleopteren von Japan mit Angabe der bezüglichen Beschreibungen und der sicher bekannten Fundorte. *Jahrbücher des Nassauischen Vereins für Naturkunde* 40: 31–204.

[1]Schrank von Paula F. 1789: Entomologische Beobachtungen. *Naturforscher* 24: 60–90.

[3]Schrank von Paula F. 1795: *Naturhistorische und ökomomische Briefe über das Donaumoor. Nebst einer Kupfertafel*. Mannheim: Schwan und Götz, [4] + 211 + [4] pp.

[2]Schrank F. von Paula F. 1798: *Fauna Boica. Durchgedachte Geschichte der in Baiern einheimischen und zahmen Thiere. Erster Band, Zweyte Abtheilung*. Nürnberg: Stein, 295–720 pp.

[1]Semenov A. P. (Tian-Shanskiy) 1891: Diagnoses Coleopterorum novorum ex Asia Centrali et Orientali. *Trudy Russkago Entomologicheskago obshchestva* (*Horae Societatis Entomologicae Rossicae*) 25 [1890–1891]: 262–382.

Silfverberg H. 1994: Case 2713. COLYDIIDAE Erichson, 1842 (Insecta, Coleoptera): proposed precedence over CERYLONIDAE Billberg, 1820 and ORTHOCERINI Blanchard, 1845 (1820); and Cerylon Latreille, 1802: proposed conservation of Lyctus histeroides Fabricius, 1792 as the type species. *Bulletin of Zoological Nomenclature* 51(1): 17–20.

Sittichaya W. 2023: *Bostrichidae of Thailand*. Thailand: Songkia: Thailand, 216 pp. (in Thai).

[1]Snellen van Vollenhoven S. C. & Sélys Longchamps M. E. de 1869: *Recherches sur la faune de Madagascar et de ses dépandances, d'après les découvertes de François P. L. Pollen et D. C. van Dam. Ouvrage dédié à S. M. Guillaume III, Roi des Pays-Bas. 5me Partie. 1re Livraison. Insectes*. Leyden: J. K. Steenhoff, 25 pp. + 2 pls.

Souza T. S. de, Trevisan H., Xavier R. L., Coimbra H. T. & Porto C. M. L. de 2022: Heterobostrychus aequalis (Waterhouse, 1884) (Coleoptera: Bostrichinae): Its interception at the Harbor of Rio de Janeiro and relevance as a quarentine pest (A1). *Arquivos do Instituto Biológico* 89: 1–5.

Spilman T. J. 1971: Bredin-Archbold-Smithsonian Biological Control of Dominica: Bostrichidae, Inopeplidae, Lagriidae, Lyctida, Lymexylonidae, Melandryidae, Monommidae, Rhipiceridae, and Rhipiphoridae (Coleoptera). *Smithsonian Contributions to Zoology* 70: 1–10.

Spilman T. J. 1982: False powder post beetles of the genus Dinoderus in North America (Coleoptera, Bostrychidae). *The Coleopterists' Bulletin* 36(2): 193–196.

[1)]Stefani-Perez T. de 1911: Una specie inedita di Bostrychus (Coleottero) dell'Eritrea. *Giornalle di Scienze Naturali et Economische* 28: 61–63.

[1)]Steinheil E. W. 1872: Symbolae ad historiam Coleopterorum Argentinae meridionalis, ossia enumerazione dei coleotteri dal prof. P. Strobel nell' Argentinia meridionale, e descrizione delle specie nuove. *Atti della Societa Italiana di Scienze Naturalle* 15 [1872]: 554–578.

[1)]Stephens J. F. 1830: *Illustrations of British Entomology; or, a Synopsis of Indigenous Insects: containing their generic and specific distinctions; with an account of their metamorphoses, times of appearance, localities, food, and economy, as far as practicable. Embellished with coloured figures of the rarer and more interesting species. Mandibulata. III*. London: Baldwin and Cradock, 374 + 6 pp. + pls. 16–19.

[3)]Sturm J. 1826: *Catalog meiner Insecten-Sammlung. Erster Theil. Käfer.* Mit 4 ausgemalten Kupfertafeln. Nürnberg: Verfassers, viii + 207 pp. + 16 pls + 2 + 4 pls.

[1)]Sturm J. 1843: *Catalog der Kaefer-sammlung von J. Sturm.* Nürnberg: Verfassers, xii + 386 pp. + 12 pls.

Teixeira É. P., Novo J. P. S. & Filho E. B. 2002: First Record of Sinoxylon anale Lesne and Sinoxylon senegalensis (Karsch) (Coleoptera: Bostrichoidea) in Brazil. *Neotropical Entomology* 31(4): 651–652.

Thurston G. S., Slater A., Nei I., Roberts J., McLachlan Hamilton K., Sweeney J. D. & Kimoto T. 2022: New Canadian and Provincial Records of Coleoptera Resulting from Annual Canadian Food Inspection Agency Surveillance for Detection of Non-Native, Potentially Invasive Forest Insects. *Insects* 2022, 13, 708. Doi.org/10.3390/insects13080708.

[1)]Thomson C. G. 1863: *Skandinaviens Coleoptera, synoptiskt bearbetade. Tom V*. Lund: Lunbergska, 340 pp.

[1)]Thomson J. 1858: *Archives Entomologiques ou recueil contenant des illustrations d'Insectes nouveaux or rares. The third page is entitled "Voyage au Gabon. Histoire naturelle des insectes et des arachnides recueillis pendant un voyage fait au Gabon en 1856 et en 1857 par M. Henry C. Deyrolle sous les auspices de MM. Le Comte de Mniszech et James Thomson précédée de l'histoire du voyage par M. James Thomson – Arachnides, par M. H. Lucas". Tome deuxième.* Paris, 469 + 1 (table des noms d'auteurs) + 1 (errata).

[2]Thunberg C. P. 1798: *D. D. Museum Naturalium Academiae Upsaliensis. Cujus Appedix. VI. Resp. J. E. Forsström*. Uppsala, 3 + 111–117 pp.

Tiaga Neto L. J., Pereira J. M., Rodrigues O. D., Lima N. L., Froio L. M. & Flechtmann C. A. H. 2022: Diversity of Scolytinae, Platypodinae (Curculionidae) and Bostrichidae in Hevea brasiliensis (Willd. Ex A. Juss) in the state Goias, Brazil. *Ciencias Florestal* (Santa Maria) 32(1): 493–503.

[1]Tournier H. 1874: Tableau synoptique des espèces européennes du genre Lyctus Fabr. *Petites Nouvelles Entomologiques* 6: 411–412.

Turnbow R. H. jr. & Thomas M. C. 2008: An annotated checklist of the Coleoptera (Insecta) of the Bahamas. *Insecta Mundi* 0034: 1–64.

[1]Van Dyke E. C. 1923: New species of Coleoptera from California. *Bulletin of the Brooklyn Entomological Society* 18(2): 37–53.

[3]Villa A. & Villa G. B. 1833: *Coleoptera europae dupleta in collectione Villa quae pro muttua commutatione offerri possunt*. Mediolani: Villa, 1–36 pp.

[1]Villa A. & Villa G. B. 1835: *Supplementum Coleopterorum europae dupletorum catalogo collectionis Villa idest species aliae, quae nunc pro mutua commutatione itidem offerri possunt; nec non emendationes aliquarum specierum in catalogo anni 1833 extantium*. Mediolani: Villa, 37–66 pp.

[1]Villa A. & Villa G. B. 1844: *Catalogo degli Insetti Coleopteri della Lombardia. In: Notizie naturali e civili sulla Lombardia 1*. Milano: Berdardoni, 416–478 pp.

Vrydagh J. M. 1948: Étude des Coléoptéres Bystrychides recueillis par M. A. Collart au Congo Belge. *Bulletin du Musée Royal d'Histoire Naturelle de Belgique* 24(45): 1–25.

Vrydagh J. M. 1951: Faune enotmologique des Bois au Congo Belge. Les Insects Bostrychides. *Bulletin de Agriculture du Congo Belge* 42: 65–90.

Vrydagh J. M. 1952a: Bostrychidae paléarctiques: le genre Scobicia Lesne. *Bulletin et Annales de la Société Royale d'Entomologie de Belgique* 88(3/4): 58–59.

Vrydagh J. M. 1952b: Bostrychidae (Coleoptera, Teredilia) de l'Angola, appartenant au Museée de Dundo (1ére note). *Publicaçõex Culturais, Companhia de Diamanthem de Angola, Srvicio Culturais, Dundo-Lunda-Angola* 15: 93–100.

Vrydagh J. M. 1953a: Contribution à l'étude des Bostrychidae (Coleoptera Teredilia). 1. – Les Bostrychides du Sahara. *Bulletin de l'Institut Royal des Sciences Naturelles de Belgique* 29(4): 1–4.

Vrydagh J. M. 1953b: Contribution à l'étude des Bostrychidae (Coleoptera Teredilia). 2. – Les Bostrychides du Soudan anglo-égyptien. *Bulletin de l'Institut Royal des Sciences Naturelles de Belgique* 29(26): 1–6.

Vrydagh J. M. 1953c: Contribution a l'etude de la zone d'inondation du Niger (mission G. Remaudiere, 1950) III. Coleoptères, Bostrychidae. *Mémoires de l'Institut Français d'Afrique Noire* 15: 1535–1538.

[1]Vrydagh J. M. 1954: *Bostrychidae (Coleoptera, Teredilia). Exploration de la Parc Naturelle Upemba, Mission de Witte (1946–1949)*. Brussels, fasc. 25: 25–43.

[1)]Vrydagh J. M. 1955a: Contribution a l'etude des Bostrychidae. Coleoptera Teredilia. 3. – Les Bostrychides du Mozambique. *Bulletin de l'Institut Royal des Sciences Naturelles de Belgique* 31(16): 1–23.

[1)]Vrydagh J. M. 1955b: Contribution a l'étude des Bostrychidae (Coleoptera Teredilia). 4. – Collection du Musée Zoologique de l'« Hunboldt-Universität » à Berlin. *Bulletin de l'Institut Royal des Sciences Naturelles de Belgique* 31(41): 1–16.

[1)]Vrydagh J. M. 1955c: Contribution a l'étude des Bostrychidae (Coleoptera Teredilia). 5. – Collection de la « California Academy of Sciences ». *Bulletin de l'Institut Royal des Sciences Naturelles de Belgique* 31(53): 1–15.

[1)]Vrydagh J. M. 1955d: Contribution a l'etude des Bostrychidae. VI. – Descriptions de Bostrychidae nouveaux. *Bulletin et Annales de la Société Royale d'Entomologie de Belgique* 91(9–10): 261–266.

Vrydagh J. M. 1956a: Contribution à l'étude des Bostrychidae. VII. Le genre Dinoderus Stephens 1830. *Mémoires de la Société Entomologique de Belgique* 27: 495–513.

Vrydagh J. M. 1956b: Contribution a l'etude des Bostrychidae. 8. – Collection de la « Zoologische Sammlung des Bayerischen Staates » à Munich. *Bulletin de l'Institut Royal des Sciences Naturelles de Belgique* 32(6): 1–20.

Vrydagh J. M. 1956c: Contribution à l'étude des Bostrychidae. 9. Le Genre Amintinus Lesne. *Bulletin & Annales de la Société Royale d'Entomologie de Belgique* 92: 64–66.

[1)]Vrydagh J. M. 1956d: Contribution à l'étude des Bostrychidae. 10. Descriptions de Bostrychidae nouveau. *Bulletin & Annales de la Société Royale d'Entomologie de Belgique* 92: 257–262.

[1)]Vrydagh J. M. 1957: Contribution a l'etude des Bostrychidae. XII. – Révision du Genre Xylobosca Lesne, 1900. *Bulletin de l'Institut Royal des Sciences Naturelles de Belgique* 33 (52): 1–23.

[1)]Vrydagh J. M. 1958a: Contribution à l'étude des Bostrychidae. 11. Les Bostrychidae de l'Australie, de la Tasmanie et la Nouvelle-Zélande. *Bulletin & Annales de la Société Royale d'Entomologie de Belgique* 94(1–2): 35–64.

Vrydagh J. M. 1958b: Contribution à l'étude des Bostrychidae. 13. Description d'Allotypes. *Bulletin & Annales de la Société Royale d'Entomologie de Belgique* 94(5–6): 156–158.

[1)]Vrydagh J. M. 1958c: Contribution à l'etude des Bostrychidae (Coleoptera). 14. – Deuxiéme collection du Musée zoologique de l'Université Humboldt à Berlin. *Bulletin de l'Institut Royal des Sciences Naturelles de Belgique* 34(38): 1–28.

Vrydagh J. M. 1958d: Contribution à l'étude des Bostrychidae. 16. – Collection du Musée G. Frey a Tutzing. *Entomologische Arbeiten aus dem Museum G. Frey Tutzing bei München* 9: 1068–1077.

Vrydagh J. M. 1958e: Contribution à l'étude des Bostrychidae. 17. – Les types de Imhoff, 1843. *Bulletin & Annales de la Société Royale d'Entomologie de Belgique* 94(11–12): 346–347.

Vrydagh J. M. 1959a: Contribution à l'étude des Bostrychidae. 18. Additions et corrections à l'étude des Bostrychides de l'Australie. *Bulletin et Annales de la Société Royale d'Entomologie de Belgique* 95(1–4): 42–46.

Vrydagh J. M. 1959b: Contribution à l'étude des Bostrychidae. 19. Nouvelles additions à l'étude des Bostrychides de l'Australie. *Bulletin et Annales de la Société Royale d'Entomologie de Belgique* 95(9–10): 274–285.

[1)]Vrydagh J. M. 1959c: Contribution à l'étude des Bostrychidae. 20. – Descriptions d'espèces nouvelles. *Bulletin de l'Institut Royal des Sciences Naturelles de Belgique* 35(42): 1–15.

Vrydagh J. M. 1959d: Coleoptera. Bostrychidae. Un aperçu des Bostrychinae, Dinoderinae et Lyctinae de l'Afrique du Sud. *South Africa Animal Life* 6: 97–124.

Vrydagh J. M. 1959e: Contribution à l'etude des Bostrychidae. 15. Collection du Centre de Faunistique de la France d'Outre-Mer à Paris. *Bulletin de l'Institut Royal des Sciences Naturelles de Belgique* 35(23): 1–8.

Vrydagh J. M. 1959f: *Coleoptera. Bostrychidae. Un aperçu des Bostrichinae, Dinoderinae et Lyctinae de l'Afrique du Sud. South Africa Animal Life, Lund University Expedition in 1950–51.* Uppsala, VI. Pp. 97–124.

[1)]Vrydagh J. M. 1960a: Contribution à l'étude des Bostrychidae. 21 – Deuxième collection de l'Académie californienne des Sciences. *Bulletin de l'Institut Royal des Sciences Nattureles de Belgique* 36(14): 1–20.

Vrydagh J. M. 1960b: Contribution à l'étude des Bostrychidae. 23 – Collection de la Section Zoologique du Musée National Hongrois à Budapest. *Bulletin de l'Institut Royal des Sciences Nattureles de Belgique* 36(39): 1–32.

[1)]Vrydagh J. M. 1961a: Contribution à l'etude des Bostrychidae. 27. – Collection du Musée d'Histoire naturelle Senckenberg à Francfort-sur-Main. *Bulletin de l'Institut Royal des Sciences Naturelles de Belgique* 37(4): 1–23.

[1)]Vrydagh J. M. 1961b: Contribution à l'étude des Bostrychidae. No 28. – Étude des types de Fåhraeus, désignation de Lectotypes. *Bulletin de l'Institut Royal des Sciences Naturelles de Belgique* 37(7): 1–10.

Vrydagh J. M. 1961c: Présence en Belgique d'un Lyctidae exotique: Trogoxylon aequale Woll. *Bulletin & Annales de la Société Royale d'Entomologie de Belgique* 97: 37–39.

Vrydagh J. M. 1961d: Contribution à l'etude des Bostrychidae. 30. – Le Genre Xylogenes Lesne. *Bulletin de l'Institut Royal des Sciences Naturelles de Belgique* 37(10): 1–11.

Vrydagh J. M. 1962: Contribution à l'étude des Bostrychidae (Coleoptera). No 31. – Troisième collection du Musée zoologique de l'Université Humboldt à Berlin. *Bulletin de la Institut Royal des Sciences Naturelles de Belgique* 38(4): 1–47.

Vrydagh J. M. 1963a: VIII. Coleoptera Bostrychidae. La Réserve naturelle intégrale du Mont Nimba. Fascicule V. *Mémoires de l'Institut Français d'Afrique Noire* 66: 269–273.

Vrydagh J. M. 1963b: *Bostrychidae (Coleoptera, Teredilia). Exploration du Parc National Albert.* (2) 16: 37–71.

Vrydagh J. M. 1965: Coleopterous Bostrychidae recolles par J. Mateu dans l'Ennedi. *Bulletin de l'Institut Français d'Afrique Noire* 27: 724–726.

[1)]Walker F. 1858a: Characters of some apparently undescribed Ceylon Insects. *The Annals and Magazine of Natural History* (3) 2: 202–209.

[1)]Walker F. 1858b: Characters of some apparently undescribed Ceylon Insects. *The Annals and Magazine of Natural History* (3) 2: 280–286.

[1), 3)]Walker F. 1859: Characters of some apparently undescribed Ceylon Insects. *The Annals and Magazine of Natural History* (3) 3: 258–265.

[1)]Walker F. 1871: *List of Coleoptera collected by J. K. Lord, Esq. In Egypt, Arabia and near the Africain shore of the Red Sea. With characters of the undescribed Species*. London: E. W. Janson, 19 pp.

[1)]Waltl J. 1832: Ueber das Sammeln exotischer Insecten. *Faunus. Zeitschrift für Zoologie und Vergleichende Anatomie* 1(3): 166–170.

[3)]Waltl J. 1839: Käfer um Passau. *Isis Encyclopädische Zeitschrift vorzüglich für Naturgeschichte, vergleichende Anatomie und Physiologie, von Oken* 32: 221–227.

Waterhouse C. O. 1876: New Species of Coleoptera from the Island of Rodriguez, collected by the Naturalists accompanying the Transit-of-Venus Expedition. *The Annals and Magazine of Natural History* (4) 18: 105–121.

[1)]Waterhouse C. O. 1879a: An Account of a small Series of Coleoptera from the Island of Johanna. *The Annals and Magazine of Natural History* (5) 3: 360–363.

Waterhouse C. O. 1879b: Coleoptera. Zoology of Rodriguez. *Transaction of the Royal Society of London, Ser. B.* 1879: 168–527.

[1)]Waterhouse C. O. 1881: On the Coleopterous Insects collected by Prof. I. Bailey Balfour in the island of Socotra. *Proceedings of the Zoological Society of London* 1881: 469–478 + pl. 43.

[1)]Waterhouse C. O. 1884: On the Coleopterous Insects collected by H. O. Forbes in the Timor-Laut Islands. *Proceedings of the Zoological Society of London* 16: 213–219.

[1)]Waterhouse C. O. 1888: Some Observations on the Coleopterous Family Bostrichidae. *The Annals and Magazine of Natural History* (6) 1: 348–350.

Waterhouse C. O. 1894: Coleoptera (partim). Pp. 64–71. In: Walker J.: A Visit to Damma Island, East Indian Archipelago. With Notes on the Fauna. *The Annals and Magazine of Natural History* (6) 14: 49–71.

[1)]Watt J. C. 1975: Notes on priority of family-group names in Coleoptera. *The Coleopterists Bulletin* 29(1): 31–34.

[1)]Weber F. 1801: *Observationes entomologicae, continentes novorum quae condidit generum characteres, et nuper detectarum specierum descriptiones*. Kiliae: Impensis Bibliopolii Academici Novi, xii + 116 + [1 (Errata)] pp.

Webster R. P., Sweeney J. D., DeMerchant I. & Turgeon M. 2012: New Coleoptera records from New Brunswick, Canada: Dermestidae, Endecatomidae, Bostrichidae, and Ptinidae. *ZooKeys* 179: 127–139.

[1])Węgrzynowicz P. & Borowski J. 2015a: A new species of Lichenophanes Lesne, 1899 (Coleoptera: Bostrichidae) from Gambia. *Annales Zoologici* 65(4): 573–578.

[1])Węgrzynowicz P. & Borowski J. 2015b: Trogoxylyctus australiensis a new genus and species of Trogoxylini Lesne, 1921 (Coleoptera, Bostrichidae, Lyctinae) from Australia. *International Letters of Natural Sciences* 49: 58–62.

[3])White A. & Butler A. G. 1846: *Insects*. Pp. 1–24. In: Richardson J. & Gray J. E.: The Zoology of the Voyage of H. M. S. Erebus & Terror under the command of Captain Sir James Clark Ross, R. N., F. R. S. during the years 1839 to 1843. London: E. W. Janson, 51 pp.

White R. E. 1982: A catalog of the Coleoptera of America north of Mexico: Family Anobiidae. *United States Department of Agriculture, Agriculture Handbook* Number 529–70: i–xi + 1–58.

[1])Wickham H. F. 1912: A report on some recent collections of fossil Coleoptera from the Miocene Shales of Florisant. *Bulletin from the Laboratories of Natural of Natural History from the State University of Iowa* 6 [1911]: 3–38 + viii pls.

[1])Wickham H. F. 1913: Fossil Coleoptera from the Wilson ranch near Florisant, Colorado. *Bulletin from the Laboratories of Natural of Natural History from the State University of Iowa* 6(4): 3–29 + vii pls.

[1])Wickham H. F. 1914: New Miocene Coleoptera from Florissant. *Bulletin of the Museum of Comparative Zoology* 58(11): 423–494 + 16 pls.

[1])Wollaston T. V. 1858: On additions to the Madeiran Coleoptera. *The Annals and Magazine of Natural History* (3) 2: 407–415.

[1])Wollaston T. V. 1860a: On addition to the Madeiran Coleoptera. *The Annals and Magazine of Natural History* (3) 5: 252–267.

[1])Wollaston T. V. 1860b: On addition to the Madeiran Coleoptera. *The Annals and Magazine of Natural History* (3) 5: 358–366.

[1])Wollaston T. V. 1862: Brief Diagnostic Characters of new Canarian Coleoptera. *The Annals and Magazine of Natural History* (3) 9: 437–442.

[1])Wollaston T. V. 1864: *Catalogue of the Coleopterous Insects of the Canaries in the collection of the British Museum*. London: printed by order of the Trustees, xiii + 648 pp.

[1])Wollaston T. V. 1865: *Coleoptera Atlantidum, being an enumeration of the coleopterous insects of the Madeiras, Salvages, and Canaries*. London: J. van Voorst, xlvii + 526 + 140 pp.

[1])Wollaston T. V. 1867: *Coleoptera Hesperidum, being an enumeration of the coleopterous insects of the Cape Verde Archipelago*. London: J. van Voorst, xxxix + 285 pp.

Wollaston T. V. 1877: *Coleoptera Sanctae-Helenae*. London: J. van Voorst, xxv + 256 pp. + 1 pl.

Wood S. L. 1969: New synonymy and records of Platypodidae and Scolytidae. (Coleoptera). *The Great Basin Naturalist* 29: 113–128.

Wood S. L. & Bright D. E. 1992a: *A Catalogue of Scolytidae and Platypodidae (Coleoptera). Part 2, V: Taxonomic Index. Volume A*. Great Basin Naturalist Memoirs, Utah 13: [2] + 1–833 pp.

Wood S. L. & Bright D. E. 1992b: *A Catalogue of Scolytidae and Platypodidae (Coleoptera). Part 2: Taxonomic Index*. Volume B. Great Basin Naturalist Memoirs, Utah 13: [2] + 834–1553 pp.

Yoshitomi H. 2021: Redescription of Melagus japonicus Chûjo (Coleoptera: Bistrichidae: Polycaoninae). *Japanese Journal of Systematic Entomology* 27(1): 69–72.

3)Zetterstedt J. W. 1828: *Fauna insectorum Lapponica. Pars 1*. Hammone: Schulz, xx + [1] + 563 pp.

Zahradník P. 2006: World catalogue of the family Endecatomidae (Coleoptera: Bostrichoidea). *Studies and Reports of District Museum Prague-East, Taxonomical Series* 2: 142–144.

Zahradník P. 2015: Ptinidae, Bostrichidae and Endecatomidae (Coleoptera: Bostrichoidea) from Aldo Olexa's collection. *Folia Heyrovskyana, Series A* 23(1): 115–139.

Zahradník P. & Háva J. 2014: Catalogue of the world genera and subgenera of the superfamilies Derodontoidea and Bostrichoidea (Coleoptera: Derodontiformia, Bostrichiformia). *Zootaxa* 3754: 301–352.

1)Zahradník P. & Háva J. 2015: Two new Stephanopachys species from Baltic amber (Coleoptera: Bostrichoidea: Bostrichidae). *Studies and Reports, Taxonomical Series* 11(2): 433–435.

1)Zahradník P. & Háva J. 2016: Sawianus ornatus gen. nov. et sp. nov. from Thailand (Coleoptera: Bostrichoidea: Bostrichidae). *Studies and Reports, Taxonomical Series* 12(1): 297–300.

1)Zhang Y.-F., Meng L.-Z. & Beaver R. A. 2022: A review of the non-lyctine powder-post beetles of Yunnan (China) with a new genus and new species (Coleoptera: Bostrichidae). *Zootaxa* 5091(5): 501–545.

1)Zoufal V. 1894: Bestimmungs-Tabelle der Bostrychidae aus Europa und angrenzenden Ländern. *Wiener Entomologische Zeitung* 13(2): 33–42.

Note: Borowski J. & Węgrzynowicz P. (2007a): mention in the chapter "List of species described in or belonging to genera: *Apate, Bostrichus, Bostrychus*, and *Lyctus*" many different species. However, after checking the primary sources, in some cases it was found that either they were described in another genus that was never associated with the family Bostrichidae, or the author of the description was another author. These names are not mentioned in the publication, but the references sources from which we drew the information are listed.

Index of Subfamilies, Tribes, Genera and Subgenera

For higher taxonomic units, currently used endings are in accordance with rules of international codes. The original name is given in bold type.

Index of Species Names

Each entry in the index consists of four parts, arranged in the following sequences:
- Name – species, subspecies, infrasubspecific name, synonym, homonym, nomen dubium, nomen oblitum, nomen nudum.
- Current valid name.
- Current valid genus – is listed without brackets. If there is a valid subgenus, it is listed in brackets, but not separated from the valid genus by a comma.
- The original genus of the description of the original described species (regardless of the validity of the species) – it is always in brackets as the last category.

All the above parts are separated by a comma. Valid species are written in bold. Some parts may be missing for some entries such as when the valid species in the original genus contains only two parts.

Species that were described within the currently valid genus are valid and contain genus and species.

Species that are currently classified in other families are marked with an asterisk (*) at the beginning of the entry, and if known, the specific species is given; in some cases only the family or subfamily is stated.

Many species in old original genera listed in some publications are in the "nomen nudum" category, and their assignment to current valid species is not possible. The abbreviation [NN] is taken as substitute for "valid species" in this case and is shown in bold; proper names cannot be used as valid species. Species marked as nomen dubium [ND] have not been used in accordance with ICZN rules for more than 50 years and have appeared for completeness only in the Borowski & Węgrzynowicz (2007) catalog and now in this publication (including some new data). Under the rules of the ICZN they are invalid names.

An unclear entry is marked with "[?]", which is also marked in bold. This species Borowski & Węgrzynowicz (2007) is listed as representative of the family Curculionidae, subfamily Scolytinae, but even Wood & Bright (1992) in their world catalog of the subfamily Scolytinae (then still family Scolytidae) do not mention these species, and they are not mentioned in the current Palearctic catalog of the superfamily Curculionoidea (Alonso-Zarazaga 2023). Borowski & Węgrzynowicz apparently based these entries on personal comparison of the type material from various museum collections, but they have not reached final determination of the species.

Other abbreviations are also listed in the index – see Material and methods.

In the text and index, the term "homonym" is often used for the expression "nomen nudum", which is not in accordance with the rules of the ICZN. However, we wanted to point out that a number of names, often by the same authors, were used repeatedly. At this time, the rules for the use of nomenclature were not established, so the authors of the original descriptions were not clearly stated. In some cases, the assignment to a certain species appeared later in the literature, or genus, but in many cases this did not happen, even though one could "guess" which taxon the name belongs to. We completely respected that in our publication and did not try to make various speculations.

Printed in the United States
by Baker & Taylor Publisher Services